MODERN METHODS FOR AFFORDABLE CLINICAL GAIT ANALYSIS

MODERN METHODS FOR AFFORDABLE CLINICAL GAIT ANALYSIS

Theories and Applications in Healthcare Systems

ANUP NANDY

Department of Computer Science and Engineering, National Institute of Technology Rourkela, Odisha, India

SAIKAT CHAKRABORTY

Department of Computer Science and Engineering, National Institute of Technology Rourkela, Odisha, India

JAYEETA CHAKRABORTY

Department of Computer Science and Engineering, National Institute of Technology Rourkela, Odisha, India

GENTIANE VENTURE

Professor, Tokyo University of Agriculture and Technology, Tokyo, Japan

ELSEVIER

ACADEMIC PRESS

An imprint of Elsevier

Academic Press is an imprint of Elsevier
125 London Wall, London EC2Y 5AS, United Kingdom
525 B Street, Suite 1650, San Diego, CA 92101, United States
50 Hampshire Street, 5th Floor, Cambridge, MA 02139, United States
The Boulevard, Langford Lane, Kidlington, Oxford OX5 1GB, United Kingdom

Notices
Knowledge and best practice in this field are constantly changing. As new research and experience broaden our understanding, changes in research methods, professional practices, or medical treatment may become necessary.

Practitioners and researchers must always rely on their own experience and knowledge in evaluating and using any information, methods, compounds, or experiments described herein. In using such information or methods they should be mindful of their own safety and the safety of others, including parties for whom they have a professional responsibility.

To the fullest extent of the law, neither the Publisher nor the authors, contributors, or editors, assume any liability for any injury and/or damage to persons or property as a matter of products liability, negligence or otherwise, or from any use or operation of any methods, products, instructions, or ideas contained in the material herein.

Library of Congress Cataloging-in-Publication Data
A catalog record for this book is available from the Library of Congress

British Library Cataloguing-in-Publication Data
A catalogue record for this book is available from the British Library

ISBN: 978-0-323-85245-6

For information on all Academic Press publications visit our website at
https://www.elsevier.com/books-and-journals

Publisher: Mara Conner
Acquisitions Editor: Carrie Bolger
Editorial Project Manager: Gabriela Capille
Production Project Manager: Prem Kumar Kaliamoorthi
Cover Designer: Christian Bilbow

Typeset by TNQ Technologies

Contents

About the authors

Dr. Anup Nandy is working as an Assistant Professor (Grade I) in Department of Computer Science and Engineering at National Institute of Technology (NIT), Rourkela. He earned his PhD from Indian Institute of Information Technology, Allahabad, in the year of 2016. His research interest includes Artificial Intelligence, Machine Learning, Human Gait Analysis, Computing Human Cognition, and Robotics. He received an Early Career Research Award from SERB, Government of India in 2017 for conducting research on "Human Cognitive State Estimation through Multimodal Gait Analysis." He received research funding for Indo-Japanese Bilateral research project, funded by DST, Government of India and JSPS, Japan, with joint collaboration of Tokyo University of Agriculture and Technology (TUAT). He received a prestigious NVIDIA GPU Grant Award in 2018 for his research on Gait Abnormality Detection using Deep Learning Techniques. He was selected as Indian Young Scientist in the thematic area of Artificial Intelligence to participate in fifth BRICS Conclave 2020 held at Chelyabinsk, Russia, from Sept 21–25, 2020. Recently, he received research grant from DST, Government of India and Ministry of Science and ICT of the Republic of Korea in February 2021 with joint collaboration of Korea Advanced Institute of Science and Technology. He has published a good number of research papers in reputed conferences and journals.

Saikat Chakraborty obtained his MTech from Jadavpur University. Currently he is a PhD research scholar in the Computer Science and Engineering Department at NIT, Rourkela. Beside human gait analysis, he has research experience of two years in machine learning in the field of video summarization and sentiment analysis. His current research interests include computational neuroscience and computational biomechanics. He also worked as a visiting researcher in GV lab, TUAT, Japan.

Jayeeta Chakraborty is a PhD scholar in the department of Computer Science and Engineering in NIT, Rourkela. Her current research interests include Machine Learning, Human Gait Analysis, Signal and Image Processing. She has previous research experience in the domain of Data Mining, Recommendation Systems, and Semantic Web.

Gentiane Venture is a French Roboticist working in academia in Tokyo. She is a distinguished professor with TUAT and a cross appointed fellow with AIST. She obtained her MSc and PhD from Ecole Centrale/University of Nantes in 2000 and 2003, respectively. She worked at CEA in 2004 and for six years at the University of Tokyo. In 2009, she started with TUAT where she has established an international research group working on human science and robotics. With her group she conducts theoretical and applied research on motion dynamics, robot control, and nonverbal communication to study the meaning of living with robots. Her work is highly interdisciplinary, collaborating with therapists, psychologists, neuroscientists, sociologists, philosophers, ergonomists, artists, and designers.

Preface

Gait analysis has now become a well-accepted standard to assess various diseases in the clinical sector. However, traditional clinical gait analysis using high-end devices associates huge cost, which is an economic burden for many clinics and rehabilitation centers, especially in developing countries. In addition, expert assistance required for mounting, calibrating, and postprocessing of data from those devices makes it infeasible for preliminary in-home gait assessment. Hence, researchers are gradually inclining toward low-cost gait analysis using some affordable and easy-to-use sensors. But, in literature, there is a lack of a concise material that provides a clear and exhaustive documentation of affordable gait assessment systems. Due to the absence of proper guideline and tutorial for experimental setup, data collection procedure, and data analysis technique for these devices, most of the clinics tend towards traditional subjective gait analysis procedure. Therefore, a comprehensive tutorial on clinical gait analysis using low-cost devices, their validity, and applicability in recent clinical practice is presented in this book. This highly demanding issue will encourage physiotherapists and rehabilitation engineers to set up a low-cost gait analysis lab.

People with neuromusculoskeletal disorders, especially from developing countries, are seeking a convenient and low-cost system for early and in-home detection of gait abnormalities. Lack of concise materials and proper guidelines makes it difficult for pathologists and clinicians to construct a gait assessment system using low-cost sensors. This book deals with this issue by providing an exhaustive and comprehensive documentation of affordable gait analysis related to patients who have been suffering from different kinds of neuro-musculoskeletal disorders. The content of this book also tries to bridge the gap between engineering and biomedical field as it diagnoses and monitors neuromusculoskeletal abnormalities using latest technologies, especially machine learning techniques. It also includes information on how an early detection technology allows us to take precautionary measures through the development of a clinical gait analysis tool.

Acknowledgment

The preparation and the completion of this book have been possible only because of the guidance, inspiration, and assistance of many people, which we are fortunate enough to receive. Firstly, we would like to thank our institution NIT, Rourkela for providing us the opportunity and laboratory to conduct the research and experiments. We are immensely grateful to Department of Science and Technology (DST), India, and Government of India for supporting our projects with financial aids to procure the required resources for various experiments through project file no: DST/INT/JSPS/P-246/2017.

We would also like to express our sincere gratitude to Indian Institute of Cerebral Palsy and Manovikas Kendra, Kolkata, for allowing us to collect data of cerebral palsy and autistic children, respectively. We are also thankful to the staff and students of these institutions for their cooperation and assistance during the data collection process. We would also like to thank GV lab, TUAT, Japan, for supporting us with different participants' gait data. We also appreciate Sparkfun and their in-house photographer Juan Pena for letting us use their photographs.

Our sincere thanks to our professors and teachers for inspiring and guiding us with helpful suggestions. We also owe a debt of profound gratitude to our fellow labmates Bhosale Yugandhara Shivaji, Sourav Chattopadhya, Harin Santosh Dabbiru, and other members for actively participating in and assisting us during the research.

We would like to thank our parents and friends for their constant support and encouragement. Lastly, we extend our gratitude to all the students, teaching and nonteaching staff of our institute and collaborating institutes, and everyone who helped us directly and indirectly.

Introduction

1.1 What is gait?

Walking is a behavioral phenomenon of animal. It falls under one of the categories of more commonly used technical term *gait*, which is generally characterized as a quasi-periodical event of loading and unloading of limbs [1,2]. However, because of popularity, walking has been used interchangeably with gait in the state-of-the-art [3].

1.2 Gait cycle

Generally, gait has been analyzed by extracting movement trajectory of body joints or muscles and segmenting the time series in multiple cycles. To estimate a cycle, the quasi-periodic property of gait is utilized. During walking, each leg goes through a sequence of repetitive steps. Traditionally, for a normal person, the starting of a cycle is marked by the heel strike of a leg and ended with the subsequent same event of the same leg (*ipsilateral*) which is also considered as the starting of the next cycle [3].

A gait cycle can be broadly divided into two phases: stance and swing. The definitions of these phases are relative to a particular lower limb. During stance phase, the foot of the corresponding limb is on the ground, whereas in swing, the foot is no longer in contact with the ground, i.e., it is swinging through to move the body forward. Considering the both limbs all together, stance phase can be further subdivided into the following four phases (see Fig. 1.1):

- **First double limb support:** It happens when both feet are on the ground during the starting period of a cycle.
- **First single limb support:** It happens during the starting period of a cycle when one limb is in contact with the ground but the other one is swinging. For example, in Fig. 1.1, the right limb is swinging, hence this phase is labeled as left single limb support.

Modern Methods for Affordable Clinical Gait Analysis. https://doi.org/10.1016/B978-0-323-85245-6.00012-6

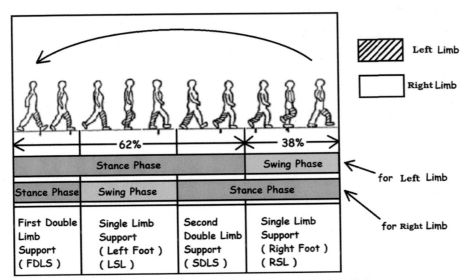

Figure 1.1 Division of a gait cycle considering both limbs.

- **Second double limb support:** It happens when both feet are on the ground again after the first single limb support.
- **Second single limb support:** It happens at the end of a cycle with same set of events like the first single limb support but with alternative limbs. For example, in Fig. 1.1, the left limb is swinging, hence this phase is labeled as right single limb support.

Generally, the phase offset between the two limbs is 50% of the cycle [4], i.e., swing phase of one limb overlaps with the mid of stance phase of the other (*contralateral*) limb. For a particular limb, the stance phase takes $\approx 62\%$ of a cycle, whereas swing phase takes the remaining [4]. Stance and swing phases can be subdivided further into five and three subphases, respectively. According to the traditional nomenclature, the stance phase events of a limb are as follows:

- **Heel strike:** Usually it represents the starting of a gait cycle.
- **Foot-flat:** It happens when the foot comes in contact with the ground with its planter surface.
- **Midstance:** This event occurs when the contralateral foot crosses the stance foot.
- **Heel-off:** It happens when the heel just get detached from the ground.
- **Toe-off:** It occurs when the foot loses the last contact with the ground. This event ends the stance phase.

The swing phase events are:
- **Acceleration:** It happens just after the toe-off event when the foot starts to accelerate in the forward direction.
- **Midswing:** It occurs when the foot passes its contralateral foot.
- **Deceleration:** It happens just before the starting of the next cycle when body muscles slow down the limb to become it stabilize again on the ground.

It is to be noted that the above-mentioned events are only for the normal gait. A more generalized version of gait events which can be attributed to any gait type was developed by Perry et al. [5]. It also consists of eight subphases namely (1) initial contact, (2) loading response, (3) midstance, (4) terminal stance, (5) preswing, (6) initial swing, (7) midswing, and (8) terminal swing.

1.3 Features of gait

Generally, gait data are represented as a time series which is restructured (after removal of noise) using some mathematical formulation to obtain a biomechanical variable or feature which is not measurable directly [6]. Salient features are the key factors to analyze gait. Features can be classified into five categories:

1.3.1 Spatio-temporal
This type of features, such as stride length, stride time, step length, step width, cadence, etc., characterizes gait variables in spatial and time domain.

1.3.2 Kinematic
This type of features relates internal dynamics of gait and characterizes the movement without referring the cause [6]. This comprises linear and angular motion parameters, i.e., displacement, velocities, and acceleration.

1.3.3 Kinetic
This contains the set of features like torque, momentum, power, etc., which describe the cause of motion [7]. Generally, features in this category are computed by applying inverse kinematic on ground reaction force (GRF) and joint kinematics [8].

1.3.4 Anthropometric

These are human body-centric features which represent physical parameters such as height, weight, age, limb length, etc.

1.3.5 Electromyography

This type of feature represents muscle activation pattern which is crucial for constructing different gait models and performing simulation.

1.4 Model-based versus model-free gait assessment

1.4.1 Model-based human gait analysis

In model-based method, a couple of approaches used different representations such as stick models, ellipsoidal models, rectangle models, and trapezoidal models for human modeling. A coveted model is fitted in each frame on person's gait sequences through this process. For recognition model, parameters are measured from walking sequences.

The human body oscillation can be depicted as several constituent segments which integrates knowledge about the dynamics and shape of human body. Therefore, it is required to investigate each part separately to extract useful information for classification. The model-based approaches provide a direction for extraction of static structural parameters of human body. A motion model can also be fitted to human body to derive joint angle trajectories of leg oscillations. Keeping in view the human anatomical structure, different articulated models can be represented for extracting prominent gait signatures. The potential advantages of model-based approaches are efficacy for handling self-occlusion, changing view, and less sensitive to noise and different apparels. The drawback of this approach is the requirement of high-resolution images with intensive computational process and affects the performance of gait recognition in outdoor environment. The analysis of previous studies focuses on fitting ellipsoidal models [9] to each portion of human body structure where model parameters are measured for each segment of human silhouette. This research work was further extended to develop a view invariant recognition system using this ellipse fitting strategy to a large gait dataset. The combination of dynamic and static features was proposed by Ref. [10] to enrich human recognition. The static information of human body was obtained by Procrustes shape analysis whereas the dynamic information was collected by fitting a motion model which retrieved joint angle trajectories of limbs oscillations. The fusion strategy was applied at a decision level to enhance the recognition result. Abdelkader et al. suggested the dynamic structural body parameters such as stride length and cadence for recognition [11]. To extract those static gait features from the subject's body, analysis of variance method was applied on the width of subject's bounding box. A simple model with three sticks has been proposed by Bobick et al. [12] to represent torso and two limbs concentrating at center of pelvis point. They have measured the structural static parameters which include distance between both feet, distance between feet and pelvis, and distance between pelvis and head. The same three stick model was proposed by Davis et al. [13] with different feature extraction techniques. Dynamic features such as time duration of double limb support and the ratio of stance phase to swing phase have been used in that research. The foot position was determined by taking the most distance silhouette pixels along the principal axis of each leg. A dynamic gait model was conceptualized by Yam et al. [14] with coupled pendulum. It was found that angular rotation of thigh and knee was extracted as gait signatures. The polynomial approximate function has been applied by Yoo et al. to calculate knee and hip angles from the human body structure [15]. A regression analysis technique has been used to derive the joint angles by exploiting the topological analysis of medical domains. A simple model with a single line was proposed by Cunado et al. for both leg oscillations to the hip [16]. The Hough transform technique was implemented to obtain the joint angles between two line segments for each silhouette frame. This research work of Ref. [16] had been extended by (Wagg and Nixon, 2004) by fitting an elliptical model on upper extremity of head and torso and represented the lower extremity of each leg using line segments. The drawback of this model includes high-computational cost for model fitting. To bring down the cost, Bouchrika and Nixon proposed another model-based approach where gait features have been extracted by manipulating the heel strike information of an individual gait pattern [17].

Another model-based approach was proposed by Cunado et al. for human gait classification [18]. The series of pendulum were used to model the legs. The thigh was modeled using the upper pendulum which connected from hip to knee. The joining of knee toward ankle was modeled by the lower pendulum. The gait signatures were obtained after applying the Fourier series analysis on angular rotation of knee and hip. Best quality image sequences are required for modeling the thigh and lower knee. This process takes more computation cost. The neural network–based gait recognition has been described by Yoo et al. using silhouette-based approach [19]. The measurement of anatomical body structure was reported to extract the two-dimensional (2D) stick model from the sequence of gait frames. The series of 2D stick figures were utilized for extracting the gait signatures using computation of motion parameters. It was observed that neural network yields a satisfactory classification accuracy with incurring high-computational cost for training network model. Consequent upon the gait analysis using the model-based approaches, the following observations are made in human gait recognition problem.

It is a very complex task to fit the correct model for each subject considering a large number of gait parameters. Therefore, a higher dimensional search space is required which leads to an increase in the computational cost, at the same time it opens the floodgates for the "curse of dimensionality" to wreak havoc with the system model.

1.4.2 Model-free human gait analysis

Human gait recognition using the model-free approaches can be done through the analysis of moving shape and motion of the subject's body. The benefit of this approach is that the recognition can be performed at large distance with sufficiently low-resolution images. This approach is very simple and intuitive to extract gait signatures from the gait frames. This model-free approach may be addressed from different view point. The model-free gait recognition techniques are not limited to background subtracted images where human silhouette from the static background is identified by measurement of silhouette shape and motion. It has been observed that original textured images, key frame analysis, stereo vision using depth sensor imaging, content-based image retrieval on still images using Kinect device were also explored to extract gait features.

The unwrapped silhouette-based approach was proposed by Wang et al. where background subtraction technique was used to extract spatial silhouette [20]. The shape analysis method was done to obtain an Eigen shape as gait signature. Sarkar et al. proposed silhouette similarity technique where the baseline algorithm was used for silhouette estimation by background subtraction and the similarity computation by temporal correlation of the silhouettes [21]. The steps of computing silhouette similarity of gait sequences include silhouette extraction, gait period detection, and spatial-temporal correlation. The human recognition was accomplished by measuring the temporal correlation of extracted silhouettes. This algorithm performed well on five covariate factors (carrying briefcase, elapsed time, different view angle, different shoe types, and different walking surface) on a dataset of 122 subjects.

BenAbdelkader et al. proposed self-similarity plot of a moving person which was a projection of its planar dynamics [22]. The self-similarity matrix encodes the frequency and phase and thus preserves the gait dynamics for recognition.

Collins et al. proposed key frame analysis technique which implicitly captures biometric shape cues such as body height, width, and body-part proportions, as well as gait cues such as stride length and amount of arm swing [23]. The key frame analysis enabled the sequence matching with innate viewpoint dependence. The key frames have been compared from the training frames using normalized correlation. The subject classification was performed by nearest neighbor matching among correlation scores.

The discrete symmetry operator used edge maps of images from the sequences of subject silhouettes [24]. It assigns symmetry magnitude and orientation to image points, accumulated at the midpoint of each analyzed pair of points. The total symmetry magnitude (also known as isotropic symmetry) of each point is the sum of the contributions of image points which have each point as their midpoint. An area-based approach was proposed in Ref. [25], where a horizontal line isolates those parts in the region of the waist. A vertical line selects the thorax and these parts of the legs intersecting with the vertical window. For each mask, gait signature was obtained by determining the area which was congruent between the mask and the images. Wang et al. proposed Eigen space sequences technique to extract gait signatures [26]. In this method, first step involved detecting and tracking the walking figure in an image sequence. In the second step, binary silhouette was extracted from each frame and convert the 2D silhouette image into a one-dimensional normalized distance signal through unwrapping the contour with respect to the silhouette centroid. Finally, the components of gait signatures are computed on those time-varying distance signals using principal component analysis. The average silhouette-based approach was proposed by Liu et al. [27]. The first step was silhouette extraction in each frame from the background pixel statistics. The second step was to estimate the gait periodicity. The third step was average silhouette computation.

Further approach of gait recognition employs the representation of Frieze patterns by exploiting the vertical and horizontal projections of human silhouettes. The temporal sequences of human gait have been modeled using a Hidden Markov Model (HMM) [28]. This HMM model was further used by Kale et al. to train the system directly using the features such as the entire silhouette image and the width of the outer boundary of a binary silhouette [29]. The silhouette shape of the subject's motion was analyzed by Little et al. for individual recognition [30]. The dense optical flow has been extracted for sequence of silhouette frames which provides the scale invariant features of each flow. This scale feature of each image sequence contains the same period with different phases used for recognition by exploiting the shape analysis. Boulgouris et al. proposed a radon transform technique applied on binary silhouette images [31]. The feature vector was derived from the radon coefficients by employing the subspace projections and applying linear discriminant analysis.

Lili Liu et al. proposed a technique to derive distance signals as feature vector from the outermost contour of silhouette images [32]. The PCA was then applied to reduce the dimensionality of feature space to enrich the classification accuracy. The high-quality silhouette images were produced using locality preservation projections method which was proposed by Wang et al. [33]. Principal component analysis method was used to reduce the dimension of the silhouette features. A regression-based view transformation model was developed by Kusakunniran et al. to remove view variations [34].

A low-cost Kinect sensor having integrated depth camera is used for recent human gait recognition with the facility of automatic extraction of intrinsic gait features. The depth map information is used to render for human skeleton detection and tracking. The joint position information is extracted from a stick model through Kinect-based gait analysis. Each joint of skeletal figure is converted into a pixel of particular color using content-based image retrieval technique on still images [35]. The human detection and classification has been extended by Gill et al. using the stereo vision techniques with Microsoft Kinect Xbox 360 with infrared depth sensor [36]. The analysis of person's walking in voxel space has been carried out with different algorithms to compare both the experimental results. Stone et al. proposed for in-home gait assessment using Kinect depth camera sensor for seven older subjects [37]. The essential gait parameters such as stride length, stride time, height, and average speed have been estimated to monitor the early fall in gait disorder problems. Kinect sensors and web camera provide facility for another gait assessment using stride to stride gait measurement method [38]. It allows in measuring several gait parameters such as stride length, stride velocity, and stride time for the prediction of gait disorder. A hybrid feature selection method was proposed by A.Sinha et al. with both static and dynamic features from a human skeleton model [39]. The three-dimensional skeletal data were acquired from low-cost Kinect sensor for gait analysis using an unsupervised K-means clustering algorithm [40]. The lower extremity parts captured by low-cost Kinect sensor was investigated and compared with IGOD suit [41] for proliferation of Kinect-based gait assistive tool [42].

Automatic gait recognition using the model-free approaches is very simple which requires only the background subtracted binary images. The poor resolution of binary silhouette image cannot lead to good recognition result because of the distortion of silhouette shape. This approach provides flaws in developing a generic segmentation algorithm for realistic environments to extract accurate human silhouette with less noise distortion.

1.5 Applications of gait analysis

Wearable sensor-based gait analysis supports several applications in clinical field. In that direction, a gait shoe can be through off with different sensors such as accelerometers, gyroscopes, force sensors, bidirectional bend sensors, pressure sensors, and electric field height sensors to collect data unobtrusively over a long period of time in any environment. This Gait Shoe can be tested on different healthy, and Parkinson disease subjects for the clinical assessment of rehabilitation and physical therapy.

Presently, gait analysis has gained an incredible acceptance to pervade all its aspects in the field of forensics and surveillance security-based applications for identification of a legitimate individual using computer vision-based techniques as well as in clinical field. Noncontact features gait has been regarded extremely useful in the field of surveillance-based security applications where an unknown subject can be identified without their prior consent or cooperation. There are some situations where subject recognition can be done from a long distance. The merits of subject's gait are extremely beneficial in this regard. The maximum information of subject's full body can be analyzed for recognition purposes. A subject's gait data can be obtained from a distance using a simple video camera where subject's full support may not be necessarily required. This advantage attracts one to the field of gait biometric which can be used in video surveillance applications.

Human gait data can be acquired for psychological studies for long time where humans are able to recognize known subjects using their gait. In medical gait research, pathological gait patterns could be generated for normal persons to detect abnormal gait patterns using those normal patterns. Other fields of gait analysis are in medicine and biomechanics. The use of gait provides a clear insight into the medical implications in automatic diagnosis of differently abled patients by understanding the normal and abnormal gait patterns. It could also be used in rehabilitation practices for the analysis of athletic performances. The biomechanics studies demonstrate that human gait provides unique characteristics if all the movements are considered during locomotion. These extensive studies will motivate computer scientists to develop computer vision—based automatic gait recognition. It may be observed that the gait information in turn will reveal medical implications in subjects with medical disorders such as neuro-musculo-skeletal disorder, osteoporosis, etc.

1.6 Clinical aspects of human gait

The locomotion of human body is result of interaction between the central nervous system (CNS), peripheral nervous system, and musculoskeletal effector system. The sequence of events for initiating movement can be stated as followed [4]:

(1) Registration and activation of the gait command in the CNS;

(2) Transmission of the gait signals to the peripheral nervous system;

(3) Contraction of muscles that develop tension;

(4) Generation of forces at, and moments across, synovial joints;

(5) Regulation of the joint forces and moments by the rigid skeletal segments based on their anthropometry;

(6) Displacement (i.e., movement) of the segments in a manner that is recognized as functional gait;

(7) Generation of GRFs.

An injury or lesion at top level of the movement chain (CNS) affects the subsequent levels. Gait assessments can be done at different levels such as analyzing muscle activities [using electromyography (EMG) sensors], movement of the segments/limbs (using vision/inertial sensors), GRFs (using pressure sensors). In gait analysis, the motion of body segments and forces during movement is captured, and inverse dynamics are applied to retrieve significant information about the cause of the motion. For example, motion sensors do not directly measure joint angles but the movement of joint positions which can help to derive the joint angles; GRF can be helpful to identify center of gravity of the human body, etc. Assessment of an individual's gait pattern may thus help identify the level of the movement chain of the injury causing a particular gait pattern. For example, gait pattern analysis can be used to differentiate between irregular gait patterns because of neurological disease and because of accident or muscular injury. Here, we will discuss some of the aspects of gait assessment which are helpful to clinicians in health-care applications.

1.6.1 Gait signal segmentation

Segmentation of gait signals is generally performed to extract gait cycles from the complete signal. One of the major reasons of doing segmentation is that the gait pattern can be better characterized with the distinguishing features of each cycles rather than the complete signal. In addition to that, dividing signals captured for longer periods into smaller segments helps in assessment during gait analysis. The approaches for segmenting can be divided into two categories in general: (A) finding periodicity of the signal—this method is used mostly where the segments are required to be of same lengths; (B) finding a specific event—this method identifies an event such as heel strike to mark the end or start of gait cycles. Identifying gait events can help in both cycle extraction as well as for obtaining significant features of a gait cycle. Detecting the events can be used to divide the cycle into smaller segments. Many studies have proposed methods such as threshold-based, machine learning tool—based methods to identify the occurrence of gait events which will be discussed in detail in Chapter 5.

1.6.2 Pathological gait detection

Gait analysis is performed to assess the presence of any abnormality in a patient's gait pattern, and if present, then what pathological condition may be the cause of the abnormality and depending on that, what medicine and exercises should be prescribed. For clinical assessment, tools and methods are being developed to help clinicians identify both the presence and cause of gait abnormality. The traditional methods that are followed in clinics are observational, questionnaire-based, or statistical-based methods. Statistical-based methods are heuristic processes where the signal values are observed from collected data and statistical characteristics are analyzed to categorize different pathological gait patterns. With the development in the domain of pattern recognition, various powerful machine learning tools have been proposed to process and learn to extract distinguishing features from images and signals. These methods have also been used effectively for gait analysis in both image and signal processing. Chapters 6 and 7 will discuss in detail how machine learning–based methods can be used to assess pathological gait pattern.

1.6.3 Injury prevention and recovery prediction

Apart from pathological gait assessment, gait analysis is also done to monitor sports activities that can help monitor the limb movements of athletes. This may guide the athletes about the correct pose that requires to be maintained for preventing from injuries. Gait pattern is also analyzed during rehabilitation period, especially for patients with neuromuscular injuries. Regular analysis of gait pattern can show the progress of the recovery process and can also be used to predict approximate time-period that may be needed for full recovery. Advanced predictive models can be trained to monitor recovery progress of a patient. This is useful for both clinicians as well as patients as this provides some idea whether the prescribed medicines and exercises are effective or some modifications or changes are required. We will be covering this aspect of gait analysis in Chapter 10.

1.7 Sensors for gait data acquisition

The sensors used for gait data acquisition are also known as motion capture sensors. The process of acquisition, the measured outputs as well as accuracy of the output differ for various sensors. The camera sensors can capture information about joint positions and angles between different limbs of body during movement, body joint information can also be derived from inertial sensors; whereas measuring GRF requires force plates or force-sensitive resistors, on the other, the muscle energy can be measured by EMG sensors which can be significant to model an individual's gait. Factors like accuracy of the sensors are not usually compromised during clinical assessment whereas for sports, daily activities monitoring, trade-off between accuracy and affordability of the device are maintained. On the other hand, using nonportable sensors can put the constraint of only indoor data collection whereas wearable or portable sensors have the advantage to be used outside gait lab. Thus, it depends greatly on the application to select which sensor to use. In this book, we will focus on the motion capture sensors that are cost-effective (see Chapter 3) as well as explore the different areas in clinical applications where these sensors can be used effectively in the following chapters.

1.8 Summary

This chapter presents a brief introduction about human gait and the significance of gait analysis in clinical aspects. The basics of gait pattern assessment are discussed which involves around different gait features and modeling techniques. Gait analysis has various applications in clinical domain. However, expensiveness of high-end motion sensors limits the availability of these clinical services only to handful portion of wider public. In recent years, owing to advancement in sensor technologies, inexpensive and portable sensors have been developed to come in aid of cost-effective gait analysis. In this book, we will focus on these affordable sensors and analyze the recent developments that are helping to make these sensors more suitable and effective for healthcare applications.

References

[1] C. Kirtley, Observational gait analysis, Clin. Gait Anal. (2006) 267−298, https://doi.org/10.1016/b978-0-443-10009-3.50019-9.
[2] O. Afsar, U. Tirnakli, N. Marwan, Recurrence Quantification Analysis at work: quasi-periodicity based interpretation of gait force profiles for patients with Parkinson disease, Sci. Rep. (2018), https://doi.org/10.1038/s41598-018-27369-2.
[3] C. Kirtley, Clinical Gait Analysis: Theory and Practice, Elsevier, 2006.
[4] C.L. Vaughan, B.L. Davis, C.O. Jeremy, et al., Dynamics of Human Gait, 1999.
[5] S.R. Simon, Gait analysis, normal and pathological function, J. Bone Jt. Surg. (1993), https://doi.org/10.2106/00004623-199303000-00027.
[6] D.A. Winter, Biomechanics and Motor Control of Human Movement, fourth ed., 2009.
[7] R.L. Huston, Fundamentals of Biomechanics, 2013.
[8] S. Chen, J. Lach, B. Lo, G.Z. Yang, Toward pervasive gait analysis with wearable sensors: a systematic review, IEEE J. Biomed. Health Inf. 20 (6) (2016) 1521−1537, https://doi.org/10.1109/JBHI.2016.2608720.
[9] L. Lee, W.E.L. Grimson, Gait analysis for recognition and classification, in: Proceedings of Fifth IEEE International Conference on Automatic Face Gesture Recognition, 2002, pp. 155−162.
[10] L. Wang, H. Ning, T. Tan, W. Hu, Fusion of static and dynamic body biometrics for gait recognition, IEEE Trans. Circ. Syst. Video Technol. 14 (2) (2004) 149−158.
[11] C. BenAbdelkader, R. Cutler, L. Davis, Stride and cadence as a biometric in automatic person identification and verification, in: Proceedings of Fifth IEEE International Conference on Automatic Face Gesture Recognition, 2002, pp. 372−377.
[12] A.F. Bobick, A.Y. Johnson, Gait recognition using static, activity-specific parameters, in: Proceedings of the 2001 IEEE Computer Society Conference on Computer Vision and Pattern Recognition, CVPR 2001, vol. 1, 2001 pp. I−I.

[13] J.W. Davis, S.R. Taylor, Analysis and recognition of walking movements, in: Object Recognition Supported by User Interaction for Service Robots, vol. 1, 2002, pp. 315–318.

[14] C. Yam, M.S. Nixon, J.N. Carter, Automated person recognition by walking and running via model-based approaches, Pattern Recognit. 37 (5) (2004) 1057–1072.

[15] J.-H. Yoo, M.S. Nixon, C.J. Harris, Model-driven statistical analysis of human gait motion, in: Proceedings. International Conference on Image Processing, vol. 1, 2002 pp. I–I.

[16] D. Cunado, M.S. Nixon, J.N. Carter, Using gait as a biometric, via phase-weighted magnitude spectra, in: International Conference on Audio-And Video-Based Biometric Person Authentication, 1997, pp. 93–102.

[17] I. Bouchrika, M.S. Nixon, Model-based feature extraction for gait analysis and recognition, in: International Conference on Computer Vision/Computer Graphics Collaboration Techniques and Applications, 2007, pp. 150–160.

[18] D. Cunado, M.S. Nixon, J.N. Carter, Automatic extraction and description of human gait models for recognition purposes, Comput. Vision Image Understanding 90 (1) (2003) 1–41.

[19] J.-H. Yoo, D. Hwang, K.-Y. Moon, M.S. Nixon, Automated human recognition by gait using neural network, in: 2008 First Workshops on Image Processing Theory, Tools and Applications, 2008, pp. 1–6.

[20] L. Wang, T. Tan, W. Hu, H. Ning, Automatic gait recognition based on statistical shape analysis, IEEE Trans. Image Process. 12 (9) (2003) 1120–1131.

[21] S. Sarkar, P.J. Phillips, Z. Liu, I.R. Vega, P. Grother, K.W. Bowyer, The humanID gait challenge problem: data sets, performance, and analysis, IEEE Trans. Pattern Anal. Mach. Intell. 27 (2) (2005) 162–177.

[22] C. BenAbdelkader, R. Cutler, H. Nanda, L. Davis, EigenGait: motion-based recognition of people using image self-similarity, in: International Conference on Audio-and Video-Based Biometric Person Authentication, 2001, pp. 284–294.

[23] R.T. Collins, R. Gross, J. Shi, Silhouette-based human identification from body shape and gait, in: Proceedings of Fifth IEEE International Conference on Automatic Face Gesture Recognition, 2002, pp. 366–371.

[24] J.B. Hayfron-Acquah, M.S. Nixon, J.N. Carter, Automatic gait recognition by symmetry analysis, Pattern Recognit. Lett. 24 (13) (2003) 2175–2183.

[25] J.P. Foster, M.S. Nixon, A. Prügel-Bennett, Automatic gait recognition using area-based metrics, Pattern Recognit. Lett. 24 (14) (2003) 2489–2497.

[26] L. Wang, T. Tan, H. Ning, W. Hu, Silhouette analysis-based gait recognition for human identification, IEEE Trans. Pattern Anal. Mach. Intell. 25 (12) (2003) 1505–1518.

[27] Z. Liu, S. Sarkar, Simplest representation yet for gait recognition: averaged silhouette, in: Proceedings of the 17th International Conference on Pattern Recognition, 2004, ICPR 2004, vol. 4, 2004, pp. 211–214.

[28] A. Sundaresan, A. RoyChowdhury, R. Chellappa, A hidden markov model based framework for recognition of humans from gait sequences, in: Proceedings. 2003 International Conference on Image Processing, 2003, ICIP 2003, vol. 2, 2003 pp. II–93.

[29] A. Kale, et al., Identification of humans using gait, IEEE Trans. Image Process. 13 (9) (2004) 1163–1173.

[30] J. Little, J. Boyd, Recognizing people by their gait: the shape of motion, Videre J. Comput. Vis. Res. 1 (2) (1998) 1–32.

[31] N. V Boulgouris, Z.X. Chi, Gait recognition using radon transform and linear discriminant analysis, IEEE Trans. Image Process. 16 (3) (2007) 731–740.

[32] L. Liu, Y. Yin, W. Qin, Y. Li, Gait recognition based on outermost contour, Int. J. Comput. Intell. Syst. 4 (5) (2011) 1090–1099.

[33] L. Wang, D. Suter, Analyzing human movements from silhouettes using manifold learning, in: 2006 IEEE International Conference on Video and Signal Based Surveillance, 2006, p. 7.

[34] W. Kusakunniran, Q. Wu, J. Zhang, H. Li, Cross-view and multi-view gait recognitions based on view transformation model using multi-layer perceptron, Pattern Recognit. Lett. 33 (7) (2012) 882–889.

[35] M. Milovanović, M. Minović, D. Starcević, New gait recognition method using Kinect stick figure and CBIR, in: 2012 20th Telecommunications Forum (TELFOR), 2012, pp. 1323–1326.

[36] T. Gill, J.M. Keller, D.T. Anderson, R.H. Luke, A system for change detection and human recognition in voxel space using the Microsoft Kinect sensor, in: 2011 IEEE Applied Imagery Pattern Recognition Workshop (AIPR), 2011, pp. 1–8.

[37] E.E. Stone, M. Skubic, Passive, in-home gait measurement using an inexpensive depth camera: initial results, in: 2012 6th International Conference on Pervasive Computing Technologies for Healthcare (PervasiveHealth) and Workshops, 2012, pp. 183–186.

[38] E.E. Stone, M. Skubic, Passive in-home measurement of stride-to-stride gait variability comparing vision and Kinect sensing, in: 2011 Annual International Conference of the IEEE Engineering in Medicine and Biology Society, 2011, pp. 6491–6494.

[39] A. Sinha, K. Chakravarty, B. Bhowmick, et al., Person identification using skeleton information from kinect, in: Proc. Intl. Conf. on Advances in Computer-Human Interactions, 2013, pp. 101–108.

[40] A. Ball, D. Rye, F. Ramos, M. Velonaki, Unsupervised clustering of people from'skeleton'data, in: Proceedings of the Seventh Annual ACM/IEEE International Conference on Human-Robot Interaction, 2012, pp. 225–226.

[41] S. Mondal, A. Nandy, A. Chakrabarti, P. Chakraborty, G.C. Nandi, A framework for synthesis of human gait oscillation using intelligent gait oscillation detector (IGOD), in: International Conference on Contemporary Computing, 2010, pp. 340–349.

[42] A. Nandy, P. Chakraborty, A new paradigm of human gait analysis with Kinect, in: 2015 Eighth International Conference on Contemporary Computing (IC3), 2015, pp. 443–448.

2

Statistics and computational intelligence in clinical gait analysis

2.1 Introduction

For decades, clinicians have used different statistical tests to assess patient's gait. After examining a patient's gait, the clinician has to report about its present status. For this purpose, the measured gait parameters or features are compared against the normal subjects using some statistical tests. Depending on those test results, the improvement of the patient can be assessed; a random subject can be marked as normal or abnormal, or sometimes the efficacy of an intervention can be determined. Clinicians adopt different statistical tests like Mann–Whitney U test, One-way or two-way analysis of variance (ANOVA), Chi-square test, effect size, intraclass correlation coefficients, Tukey's honestly significant difference test, etc., for decision making. But, with high-dimensional gait data, statistical tests often fail to characterize gait patterns [1]. Recently, computational intelligence (CI) techniques have become very popular because of their ability to model high-dimensional nonlinear relationships. It has demonstrated precise models using different automated feature learning algorithms. Clinics are now gradually moving toward the CI-based techniques to assess gait.

2.2 Statistics in clinical gait data

This section provides a brief discussion on some statistical parameters that are commonly used to assess clinical gait data.

Modern Methods for Affordable Clinical Gait Analysis. https://doi.org/10.1016/B978-0-323-85245-6.00009-6

2.2.1 Confidence interval, p-value, and effect size

Confidence interval estimates a range of plausible values of a target parameter with an association of a confidence factor that the true value lays within that interval. *P*-value can be stated as evidence against a null hypothesis, whereas effect size reflects the strength of relationship between two variables [2]. These parameters are used frequently in clinical gait studies. For example, the statistical significance of the difference between two populations in terms of a gait feature can be expressed by *P*-value [3–5]. Generally, $P < .05$ or $P < .01$ is taken as a standard to measure the significance. It reflects whether the difference is true or it is for any errors. Effect size is generally used where the magnitude of difference between two groups is more important than simple difference [6]. It is also used in meta-analysis.

2.2.2 Statistics in clinical trials

Studies have used different statistical tests to find the relation between two gait variables of two populations. Results are used to label a class to a subject or finding significantly degraded or improved gait parameters. For example, Chang et al. [7] used independent t-test to compute the differences between gait variables of cerebral palsy (CP) and normal subjects. Kiernan et al. [8] have performed one-way ANOVA test to measure the difference between CP and normal populations' continuous gait variables. Some studies have used higher-order statistics like skewness and kurtosis, Shapiro–Wilk test, etc., to test the normality of gait data [9,10]. Statistical tests in clinical trials reveal the significance of a gait variable to differentiate two populations.

2.2.3 Systematic review and meta-analysis

Some researches have been performed to synthesize data collected from different studies using a statistical technique called *meta-analysis*. First, data are collected in a predefined systematic way, such that if anyone follows the exact procedure, he/she will obtain the same set of studies. Then for each study, the effect size is computed. The consistency of effect-sizes of individual studies is measured in meta-analysis. A summary effect is reported considering the variance of reported effects. For example, Chakraborty et al. [11] performed a meta-analysis to determine the most degraded gait parameters and dynamic stability variables in the CP population. Forest plots in Figs. 2.1 and 2.2 show some of the results of Chakraborty et al. [11]. The effect

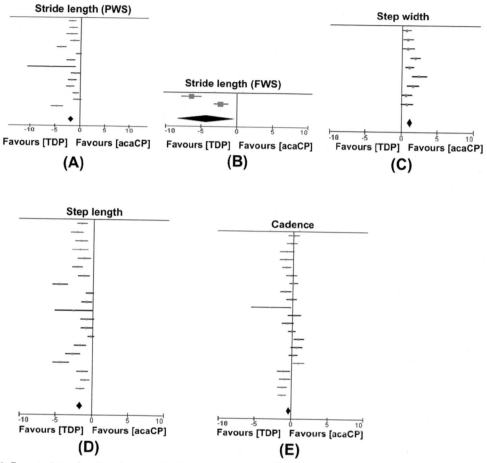

Figure 2.1 Forest plots showing the standardized mean difference (95% confidence interval) for spatial gait parameters: (A) stride length (preferred walking speed), (B) stride length (fast-walking speed), (C) step width, (D) step length, and (E) cadence. *acaCP*, ambulatory children and adolescents with CP; *FWS*, fast walking speed; *PWS*, preferred walking speed; *TDP*, typically developed children [11].

size of each study is represented by a green square. The width of the square reflects the study weight (in terms of sample size). Each line represents the range of confidence interval. The black diamond represents the summary effect. The position of this diamond determines the conclusion for that specific variable. For example, Fig. 2.1 (A) indicates that the normal population has a comparatively longer stride length than the CP population. Comber et al. [12] have also performed a meta-analysis on multiple sclerosis patients to obtain the degraded gait parameters. For further reading on meta-analysis, please see Ref. [2].

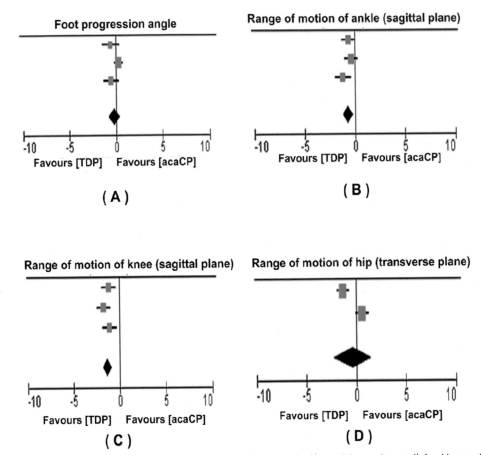

Figure 2.2 Forest plots showing the standardized mean difference (95% confidence interval) for kinematic gait parameters: (A) foot progression angle, (B) range of motion of ankle (sagittal plane), (C) range of motion of knee (sagittal plane), and (D) range of motion of hip (transverse plane) [11].

2.3 Computational intelligence in clinical gait data

Computational Intelligence (CI)-based methods for gait analysis are gradually becoming popular in the clinical domain. This section discusses some generic aspects of these techniques.

2.3.1 Why computational intelligence is important?

CI is a fusion of computing and learning mechanisms dedicated to processing and interpreting a large volume of data in decision-making systems [1]. It includes supervised learning, fuzzy classifiers, and evolutionary optimization methods. These techniques facilitate automatic feature mapping to model high-dimensional nonlinear data, making them suitable for assessing gait data.

2.3.2 Learning paradigm

The primary motivation of using CI in gait analysis is to model a biomechanical system f(x) by learning data relationships between inputs and outputs possibly corrupted by external noise $\eta(t)$ (see Fig. 2.3).

The model depicted in Fig. 2.3 can be used to detect gait abnormality, gait events, or predict the successive phase of a gait cycle. For details of the CI paradigm, see Chapter 6.

2.3.3 Applications of computational intelligence in clinical gait data

Popularity of CI methods has propelled the researchers to use it for different gait applications ranging from gait abnormality detection, event detection to prosthetic control, prosthetic control, etc. (see Fig. 2.4).

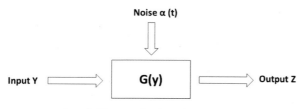

Figure 2.3 Generic learning structure of CI-based methods in gait analysis [1].

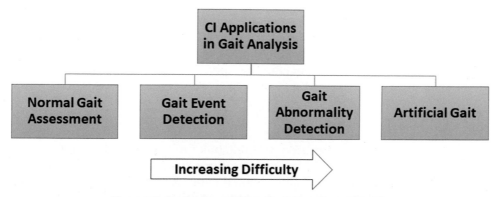

Figure 2.4 Application of CI methods in gait analysis [1].

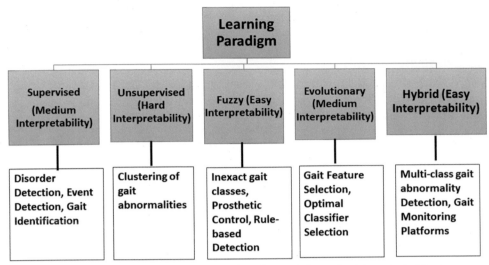

Figure 2.5 Application of different learning techniques in gait analysis [1].

Fig. 2.5 describes the most frequent application domains for different CI techniques in gait analysis. Interpretability refers to how relevant the CI outputs are to the observed biomechanical systems, with the easiest being fuzzy techniques and unsupervised learning the hardest since learning objectives have to be properly defined. The reliability of these techniques is application-dependent and usually measured by accuracy on a test set or correlation with expert opinion. For a detailed analysis of different gait application domains, please see Ref. [1].

2.4 Statistics versus computational intelligence

Both statistical methods and CI have been used extensively in clinical gait assessment. Both of them have their pros and cons. Interpretability of the statistical tests is comparatively higher. Clinicians can choose a specific statistical method depending on the extracted gait data type. Sometimes results can be obtained by simple computation, even using pen and pencil. However, as the dimension of the system increases, computation with these statistics becomes challenging and complex. Along with that, the precision and reliability of the results degrade. On the other hand, CI-based algorithms have an excellent capability to deal with high-dimensional data. They provide more robust, efficient, and cost-effective diagnostic, monitoring, and control systems by modeling the nonlinear relationship between input and output variables. Moreover, these models learn the data pattern automatically by optimizing the error function. The preciseness of these models is making them more attractive and popular in the clinical gait domain. But, the complex relationship established by these models internally is hard to interpret. Sometimes these models are referred to as a *black box*. Hence, if interpretability is more important for a clinical test, clinicians can go for the statistical investigation. But, if precision, accuracy, and reliability are major issues, then CI-based models are preferred.

2.5 Summary

This chapter describes the prevailing measurement techniques for gait assessment. Commonly used statistical metrics are briefly described. A general paradigm of CI methods has been outlined. Finally, a comparison between the statistical and CI-based techniques and a recommendation for the clinicians is provided.

References

[1] D.T.H. Lai, R.K. Begg, M. Palaniswami, Computational intelligence in gait research: a perspective on current applications and future challenges, IEEE Trans. Inf. Technol. Biomed. 13 (5) (2009) 687−702, https://doi.org/10.1109/TITB.2009.2022913.

[2] M. Borenstein, L. V Hedges, J. P. T. Higgins, and H. R. Rothstein, Introduction to Meta-Analysis.

[3] K. Desloovere, C. Motion, Motor Function Following Multilevel Botulinum Toxin Type A Treatment in Children with Cerebral Palsy, 2007, pp. 56−61.

[4] A.M.K. Wong, C.-L. Chen, C.P.C. Chen, S.-W. Chou, C.-Y. Chung, M.J.L. Chen, Clinical effects of botulinum toxin A and phenol block on gait in children with cerebral palsy, Am. J. Phys. Med. Rehabil. 83 (4) (2004) 284−291.

[5] A.M.K. Wong, Y.-C. Pei, T.-N. Lui, C.-L. Chen, C.-M. Wang, C.-Y. Chung, Comparison between botulinum toxin type A injection and selective posterior rhizotomy in improving gait performance in children with cerebral palsy, J. Neurosurg. Pediatr. 102 (4) (2005) 385−389.

[6] S.A. Radtka, S.R. Skinner, D.M. Dixon, M.E. Johanson, A comparison of gait with solid, dynamic, and no ankle-foot orthoses in children with spastic cerebral palsy, Phys. Ther. (1997), https://doi.org/10.1093/ptj/77.4.395.

[7] C.F. Chang, et al., Balance control during level walking in children with spastic diplegic cerebral palsy, Biomed. Eng. Appl. Basis Commun. 23 (6) (2011) 509−517, https://doi.org/10.1142/s1016237211002682.

[8] D. Kiernan, M. Walsh, R. O'sullivan, T. O'brien, C.K. Simms, The influence of estimated body segment parameters on predicted joint kinetics during diplegic cerebral palsy gait, J. Biomech. 47 (1) (2014) 284−288.

[9] A. Malone, D. Kiernan, H. French, V. Saunders, T. O'Brien, Do children with cerebral palsy change their gait when walking over uneven ground? Gait Posture 41 (2) (2015) 716–721.

[10] P. Meyns, L. Van Gestel, S.M. Bruijn, K. Desloovere, S.P. Swinnen, J. Duysens, Is interlimb coordination during walking preserved in children with cerebral palsy? Res. Dev. Disabil. 33 (5) (2012) 1418–1428.

[11] S. Chakraborty, A. Nandy, T.M. Kesar, Gait deficits and dynamic stability in children and adolescents with cerebral palsy: a systematic review and meta-analysis, Clin. BioMech. (2020), https://doi.org/10.1016/j.clinbiomech.2019.09.005.

[12] L. Comber, R. Galvin, S. Coote, Gait deficits in people with multiple sclerosis: a systematic review and meta-analysis, Gait Posture 51 (2017) 25–35, https://doi.org/10.1016/j.gaitpost.2016.09.026.

Low-cost sensors for gait analysis

3.1 Introduction

Human gait analysis for clinical purpose requires subtle assessment of body joint time series. Pattern analysis of joint movement is crucial in clinical gait diagnosis. Motion sensors play a prime role in acquisition of joint data. Existing motion analysis system comprising of high-end cameras provides precise results. However, those systems are very expensive and unaffordable for most of the clinics, especially in developing countries [1]. Its operating procedure requires expertise in specialized domain. Hence, as an alternative, researches are now propagating toward low-cost device-based gait analysis. Various affordable sensors, both vision-based and wearable, have been used in different studies. This chapter provides an insight on some low-cost motion sensors.

3.2 Motion capture sensors for gait

3.2.1 Classification of sensors

Motion sensors can be broadly classified as wearable and non-wearable. Both types of sensors have been used extensively in clinical gait analysis. Fig. 3.1 demonstrates the classification of different types of motion sensors.

3.2.2 Gold standard sensors

Some of the motion sensors are considered as the gold standard in different gait applications for their high accuracy, precision, and reliability. High-end vision-based cameras of different companies such as Vicon, Qualisys, etc., (see Fig. 3.2) are generally considered as the gold standard for measuring kinematic and spatio-temporal (SPT) features in gait analysis. These cameras are used to validate the output of other low-cost sensors.

Modern Methods for Affordable Clinical Gait Analysis. https://doi.org/10.1016/B978-0-323-85245-6.00008-4

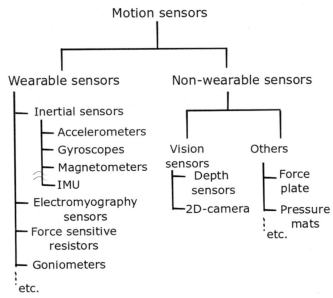

Figure 3.1 Classification of motion sensors.

Vicon (vero v1.3x)

Qualisys (Oqus)

Figure 3.2 Gold standard cameras.

For measuring kinetic features, traditional force plates, pressure mats of companies like Kistler [2–5], Tekscan (Matscan) [6–8] are considered as gold standard pressure sensors. The wearable alternative of pressure sensors is force sensitive resistors (FSRs) sensors which are embedded in shoe/sole, often referred as smart-shoe. The "F-scan system" by "Tekscan" [5,9–11] and "Pedar" by "Novel" [12–15] has been verified to be reliable and used as reference in measuring kinetic features [Ground reaction force (GRF)] during gait analysis. These sensors are generally used to validate the identification of different gait events using vision or other wearable sensors. Similar to other wearable sensors, inertial sensors are also not considered gold standard clinical gait analysis as it is less accurate in measuring the kinematic features compared to three-dimensional (3D) motion capture systems. However, among different varieties of inertial sensors available in the market, Xsens inertial measurement unit (IMU) sensors (MVN link and MVN Awinda) are considered standard for gait applications. Xsens IMU sensors come with certain advantages such as latest microelectromechanical system (MEMS) technology, easy to set-up and calibrate, software to visualize the captured data in real time, etc (https://www.xsens.com/products/mvn-analyze).

3.2.3 Affordable sensors

The gold standard motion cameras are highly expensive and non-affordable for most of the clinics, specially in developing countries. Kinect sensor has become a popular vision-based depth sensor in gait analysis because of its affordability compared to high-end cameras. This sensor has provided promising and competing results for different clinical gait assessment objectives. The wearable sensors such as inertial sensors, electromyography sensors have also gained considerable attention as affordable alternatives to traditional gait analysis systems such as 3D motion capture systems. The price range of wearable sensors also varies depending upon the type of product and the additional services related to the product. There exist companies like Sparkfun, WitMotion, etc., which manufacture significantly low-cost IMU sensors that can be used for a cost-effective gait analysis system. The cost analysis of these products is later discussed in detail in this chapter.

3.3 Microsoft kinect

3.3.1 First and second generation

Kinect version 1 (termed as *v1*) camera (see Fig. 3.3A) was first launched in 2010 basically focusing on the gaming console Xbox 360 [16] and subsequently in 2011, a software development kit (SDK) of Kinect for windows was released which facilitates real-time tracking of body joints. After gaining enormous popularity, Microsoft released second generation of Kinect (termed as *v2*) (see Fig. 3.3B) for gaming and windows in 2013 and 2014, respectively. Hardware and some software features were remarkably improved in the second generation.

(a)   (b)

Figure 3.3 (A) Kinect v1, (B) Kinect v2.

3.3.2 Hardware and software specification

Hardware: Both version of Kinect can track human body. Depth mapping is the prime feature of Kinect. Generally, two different types of depth mapping techniques are followed in Kinect: focused and stereo. Focus refers to the fact that when the distance between an object and camera increases, the object becomes blurrier. However, astigmatic lenses help to improve the tracking accuracy of Kinect [17]. Theses lenses use different focal lengths on the x and y axes. Stereo technology helps to compute depth from the disparity [17]. Kinect also has a set of microphones having noise reduction property. The microphones can also track the sound position. Kinect v1 consists of 64 MB DDR2 SDRAM and PrimeSense PS1080-A2 chip to process data [17]. The infrared emitter, having 60 mW laser diode, works on a wavelength of 830 nm. Whereas, Kinect v2 has 128 MB DDR3 SDRAM and Microsoft X871141-001 chip [17]. Theoretically frame rate for both Kinects is approximately 30 Hz, however, in practical case, many external factors like computer speed, temperature, etc., can affect the frame rate. The resolution of the color and depth sensors of Kinect v1 is 640 × 480 pixels and 320 × 240 pixels, respectively. Whereas, Kinect v2 has comparatively higher resolution (color sensor: 1920 × 1080 pixels, depth sensor: 512 × 424 pixels). The two Kinects differ in their depth mapping method. Kinect v1 estimates the distortion created by infrared dots to compute the distance, while Kinect v2 uses a Time-of-Flight (ToF) method for that purpose [17]. In Kinect v1, 3D position of a joint is obtained using stereo triangulation which provides a robust 3D reconstruction [16]. A dense depth map is obtained using surface interpolation on acquired depth values. ToF uses phase-shift distance of infrared light to compute the depth. Thus, the distance between the object and lens becomes proportional to the intensity of light [16]. This technology inherently provides the dense depth map. However, the quality of the depth image depends on the scene geometry and reflectance properties of concerned materials. A well-known problem of ToF technology is *flying pixels*. It is defined as an inaccurate estimation of depth on image boundaries [18]. Both

Kinects suffer from multipath interference problem. This happens when a pixel receives light reflected from a surface which was originally destined for other pixels [18]. For both Kinects, depth accuracy decreases as the distance between the camera and object increases [16]. View angle also creates an impact on depth estimation [19].

Software: SDK of Kinect uses the depth data to estimate the body parts based on random decision forest. The algorithm was proposed by Shotton et al. [20] where body parts were estimated using a large cohort of training set of depth images. The skeleton data stream produced by SDK after processing the depth image provides real-time body tracking based on individual joints. Kinect v1 can track up to two users while Kinect v2 can 6. Kinect v2 can track five more joint positions (i.e., 25) than Kinect v1. A comparison between Kinect v1 and v2 based on their technical features is presented in Table 3.1.

3.3.3 Data streams of kinect

Kinect provides five different data streams, i.e., color (1920×1080@30 Hz), infrared (512×424 @ 30 Hz), depth image (512×424 @ 30 Hz), body index image (512×424 @ 30 Hz), and skeleton image (20 joints @ 30 Hz for Kinect v1, and 25 joints @ 30 Hz for Kinect v2). These data streams can be accessed using SDK. Each data frame associates time stamps of local computer. Each Kinect sensor has to be connected to a dedicated computer having USB 3.0 connector. SDK can also be used to convert depth image to 3D point cloud.

3.3.4 Application in clinical gait assessment

Being a comparatively low-cost sensor, Kinect (both v1 and v2) have been used extensively for clinical gait assessment. In addition, portability of this sensor helped the clinicians and researchers to perform experiment for different sensor orientations. Its nonintrusive property makes it popular for different clinical aspects like rehabilitation [21], gait event detection [22], gait abnormality detection [23−29], monitoring activities in daily life [30], and assistive device [31] for numerous neuromusculoskeletal disorders. Some of these applications will be discussed in details in the subsequent chapters.

Table 3.1 Kinect v1 compared to Kinect v2 with their technical specifications [17].

Feature	Kinect v1	Kinect v2
Dimensions	27.94 cm × 6.35 cm × 3.81 cm	24.9 cm × 6.6 cm × 6.7 cm
Color resolution and fps	640 ×380 at 30 fps or 1280 ×720 at 12 fps	1920 × 1080 at 30 fps
IR resolution and fps	640 × 480 at 30 fps	512 × 424 at 30 fps
Depth resolution and fps	320 × 240 at 30 fps	512 × 424 at 30 fps
Field of view wide-angle lens	57° horizontal, 43° vertical	70° horizontal, 60° vertical
Specified min. distance	0.4 m or 0.8 m	0.5 m
Recommended min. distance	1.8 m	1.4 m
Tested min. distance	1 m	0.7 m
Specified max. distance	4 m	4.5 m
Tested max. distance	6 m	4 m
Active infrared	Not available	Available
Measurement method	Infrared structured light	Time of flight
Minimum latency	102 ms	20 ms
Microphone array	4 microphones, 16 kHz	4 microphones, 48 kHz
Tilt-motor	Available, ±27°	Not available
Temperature	Weak correlation	Strong correlation
More distance	Less accuracy	Same accuracy
Striped depth image	Increases with depth	No stripes on image
Depth precision	Higher	Less
Flying pixels	Not present	Present if surface is not flat
Environment color	Depth estimation unaffected	Affects depth estimation
Multipath interference	Not present	Present
Angles affect precision	No	No
Precision decreasing	Second order polynomial	No math behavior

3.4 Wearable sensors

A wearable sensor can be worn around different body parts such as wrist, waist, thigh, ankle, etc. Wearable sensors are sometimes preferred over standard laboratory-based sensors because of their portability and easy installation [32]. Thus, wearable sensors are ideal to capture motion in outdoor activities. The number of gait cycles observed per trial is also not limited by any space constraint for these sensors. There exist various types of wearable sensors that are used for gait analysis such as inertial sensors, electromyography sensors, force resistive/sensitive sensors, goniometers, inclinometers, strain gauges, etc. Depending upon the placement of the sensor, the obtained signals may vary in pattern.

Therefore, it is important that the sensors are placed correctly during experimental analysis. A detailed discussion about different wearable sensors and their applicability in low-cost gait analysis is presented next.

3.4.1 Inertial sensors

Inertial sensors work based on the principle of inertia where a body is resistant to motion. In such sensors, a proof mass on a spring is used where the proof mass deflects from its neutral position under the influence of external motion. The deflection of the proof mass is measured to record acceleration, angular velocity, etc. There are three types of inertial sensors available, i.e., gyroscopes, accelerometers, and magnetometers. A one-axle inertial sensor can measure the deflection to a single axis only whereas triaxial sensors can measure the values along pitch, yaw, roll axis. With the advancement in sensors technology, MEMSs technology is used to build the inertial sensors tiny and compact. This makes them more convenient to be worn by the participants, especially when in motion. In the following sections, we explain the working principle and general applications of the aforementioned inertial sensors built with MEMS technology.

3.4.1.1 Types of inertial sensors

(a) Accelerometer: This sensor is used to measure linear acceleration of a moving body. In a MEMS accelerometer, the proof mass is attached to a spring over a plate so that the mass can move in linear direction as shown in Fig. 3.4A. The value is measured by the change in capacitance between the movable mass and the fixed plate because of movement of the mass. Accelerometer is used to monitor vibration in machines, buildings, etc. It also finds its application in finding dynamic distance and speed in moving vehicles. It is also used to find gravitational force during aviation to monitor and control stress in aircrafts.

(b) Gyroscope: This sensor is used to measure angular velocity of a moving body. The types of gyroscope sensors available are ring laser, fiber-optic dynamically tuned, and MEMS [33]. Among different build technologies, MEMS is space and cost-efficient with adequate performance. In a MEMS gyroscope, the proof mass is also attached to springs over a plate, called sense and drive springs. However, the mass is allowed to have perpendicular displacement as shown in Fig. 3.4B. The angular velocity is calculated by change in capacitance

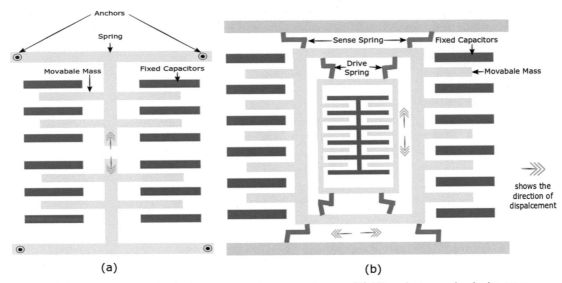

Figure 3.4 (A) Microelectromechanical system accelerometer diagram (B) Microelectromechanical system gyroscope diagram.

between the movable mass and the fixed plate, similar to accelerometer.

(c) Magnetometer: This sensor is used to measure relative change in magnetic field in a moving body. A MEMS magnetometer can be designed based on different principles such as Hall effect, Magneto resistive, Lorentz force, Electron Tunneling, etc. In Hall effect-based magnetometers, electrons are passed through a metallic plate with magnetic field nearby. The magnetic force changes the electron density distribution which creates Hall effect to generate voltage. In magneto resistive–based magnetometers, a metal sensitive to magnetic fields is used to observe the change in resistance during current flow. Apart from measuring Earth's magnetic field, magnetometer is used for understanding directions (compass) in airplanes, watercrafts, mobile phones, etc. It is also sometimes used to detect ferrous metals.

(d) Inertial measurement unit: An IMU generally consists of a tri-axle accelerometer, a tri-axle gyroscope, and a tri-axle magnetometer. Gyroscopes and accelerometers can measure only relative orientation. As magnetometer is capable of reading zero-drift orientation, it is better to use IMU to get absolute orientation and position. Combining information from the all three sensors, IMUs can be used for navigation, especially for aerial navigation and balancing. Thus, the methodology for the fusion of signals plays a great role in the effectiveness of IMU sensors. It has application in gaming to track user's head movement for virtual or augmented reality systems, tracking the motion of users handling mobile phones or remotes for game applications, etc.

3.4.1.2 Cost analysis of inertial sensors

Inertial sensors that are available in the market come with a wide range of options. The price varies depending on the build technology, accuracy, size, assisting soft-wares of the sensors. The accelerometers can be AC or DC input-based which should be selected depending on the application. Among the DC accelerometers, piezoresistive sensors are costly but more accurate than MEMS accelerometer sensors. The high-end gyroscope sensors are ring-laser or mechanical which are generally used in aerospace industry or military applications. These sensors are very large in size and accurate. The gyroscope sensors with fiber-optic technology come in the middle range which are used in race cars, torpedoes, or aircrafts. The sensors built with MEMS technologies are small and comparatively less expensive than other sensors. Thus, these are mostly used for various consumer-based applications such as in automobile, industrial, health sectors. Among the MEMS-based sensors also, it is difficult to specify a strict range boundary because there exists a number of companies which manufacture inertial sensors. Most of them have also products of wide price range to select from, depending on the application. They are generally divided as gold standard and platinum standard. However, we will broadly divide the MEMS inertial sensors used for gait analysis into two categories: high ranged (2000 dollar or above) and low ranged (100–1000 dollar). Please note that this does not reflect the exact pricing, but only to provide an approximate idea to the readers. The range may differ depending on the product specifications and customization. One of the most standard IMU sensors in literature is Xsens which belongs to the high-range class of IMU sensors [34–37]. The product kit comes with a complete setup that includes multiple IMU sensors, full body strap set, recording and docking station, dongle, UI software for the users to monitor and analyze the captured data in real-time or offline. Xsens sensors have been the most popular IMU sensor in gait studies because of its reliability, thus considered as the most standard as discussed in Section 3.2.2. Among other IMU sensors, Gaitup is an emerging company which provides a seven-dimensional inertial sensor with 3D accelerometer, 3D gyroscope, and one barometric sensor, called *Physilog 5*, which comes under low range (~$480). However, the complete kit for setting up has multiple *Physilog 5* sensors and desktop UI support that may cost more than 7000—8000 dollars reaching high-range sensor prices. Some research studies have validated Physilog for gait analysis, but mostly from the same researchers' group [38—40]. Delsys Inc provides Trigno that consists of accelerometer, gyroscopes along with Electromyography sensors as a gait monitoring kit available in high-cost range. On the other hand, research studies aiming toward low-cost gait analysis prefer self-manufactured sensors or simple inertial sensors that does not come with additional facilities such as easy-calibration, wireless control, or software for real-time monitoring. Manufacturing companies such as "Sparkfun," "WitMotion," and "Freescale Semiconductor" (ZStar3 accelerometer) markets a wide variety of MEMS IMU sensors that are well within the low range. Some of the research studies have used Sparkfun IMU sensors aiming toward low-cost gait analysis [41—43]. Some researchers use self-made or custom-made IMU sensors for their experiments providing validation of the sensors as well [44—46]. With the advancement in sensor technology, recently many research studies made attempts to model gait pattern with a single sensor present in smart-watch or smart-phone which can contribute highly to nonexpensive gait assessment systems [47,48]. Images of different inertial sensors provided by the manufacturer or developer are presented in Fig. 3.5.

3.4.1.3 Applications of inertial sensor in clinical gait assessment

In a traditional gait lab where 3D motion sensors are used, the gait analysis is done by capturing the joint angles formed by different body limbs such as knee angle, ankle angle, etc. These joint angles are then analyzed, mapped with the corresponding gait phases which are generally done using pressure mats. The measurement of joint angles along with identified phases is then used to extract SPT features. Based on these features, clinicians can give remarks about the gait pattern of a subject, whether the subject requires any treatment or whether any improvement or changes is required in the recovery program. During the early stages, a similar approach is followed for IMU sensors as well because researchers focused to use inertial sensors as a nonexpensive substitute to 3D motion capture systems in the gait lab. Thus, joint (flexion/extension) angle estimation using IMU sensors has been one of the earliest challenges and still remains a challenge especially for low-cost sensors.

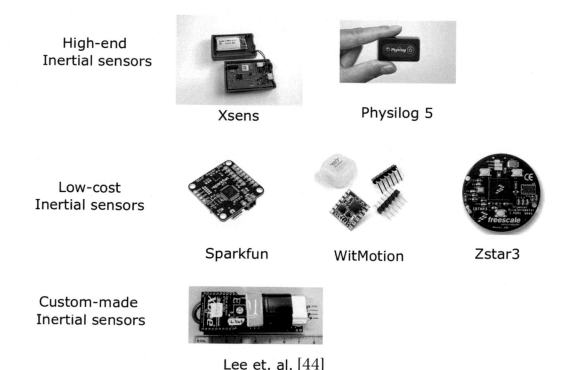

Figure 3.5 Inertial sensors belonging to different categories.

Estimation of joint angle using IMU sensor is generally validated and verified with respect to optical reference systems. To measure the flexion/extension angle from gyroscopes, integration of the change in angular rates based on the joint axis is performed. Using accelerometer measurements with respect to joint axis, the angle created between the displacement vector and the joint plane can be used to measure the joint angle. In IMU sensor-based joint angle estimation, the joint angles obtained from both accelerometer and gyroscope sensors are combined to estimate the resultant joint angle using Kalman filter [49–51] or complimentary filter [52]. A thorough explanation of the methods and steps for joint angle estimation using inertial sensors will be discussed in detail in Chapter 4.

Identification of the gait events is an important step for gait feature extraction as in conventional gait labs. In this process, the occurrence of the major events in a gait cycle are identified. Some researchers term event identification as phase detection because the algorithm detects the transition point from stance to swing and vice versa. This helps in measuring the step length, stride length, cadence, etc., features of a gait signal. Most of the early methods for event identification using IMU sensors are based on thresholding. Depending on the prespecified threshold values, the gait events are classified using the sensor observations recorded using IMU sensors. Machine learning techniques such as hidden Markov model (HMM) [53,54], maximum aposterior probability [55,56], and random forest [56] are also used for phase/event detection. A more detailed discussion on gait event detection is presented in Chapter 5.

In clinical gait analysis, a lot of efforts are being made to automate the diagnosis process of a patient. Finding out the causes to abnormal gait pattern is considered significant for early recognition of neurological diseases or identification of the presence of Musculo-skeletal injuries that can lead to unhealthy gait pattern. As opposed to the early methods of classifying depending on only statistical analysis of different gait features, with the help of machine learning techniques such as random forest [47], support vector machines [57,58], HMM [57] etc., the patterns are modeled to fit features of different disease or injury categories. In these classification techniques, many user-crafted features such as frequency, energy features, etc [57,58], are also explored apart from classic SPT features [47,48]. Recent advancement in deep learning techniques encourages automated feature extraction rather than hand-crafted features. Many studies have used raw signals [59], as well as spectrogram or FFT features [60,61], discrete wavelet transform [62], continuous wavelet transform [63] of the signal for learning features using deep learning techniques to classify pathological gait patterns. Apart from these, inertial sensors are also being used for different clinical applications such as improved controlling of prosthetic limbs during gait activities, prediction of recovery of injuries, sports/exercise activity monitoring, etc.

3.4.2 Electromyography sensors

Electromyogram (EMG) signals offer muscle activation pattern during various motor action. It is an electrodiagnostic medicine technique for evaluating and recording the electrical activity produced by skeletal muscles. EMG is performed using an instrument called an electromyograph to produce a record called an EMG. Dynamic EMG offers a means of directly tracking muscle activity. The myo-electric signal sufficiently parallels the intensity of muscle action to serve as a useful indicator of its mechanical effect. Amplitude of EMG signals derived during gait can be interpreted as a measure of relative muscle tension.

EMG is a method for assessing and recording the electrical movement delivered by skeletal muscles. EMG is performed by utilizing a device called an electromyograph to create an EMG. The EMG data are said to be proportionate with the amount of tension in the muscles. Surface EMG (sEMG) is a noninvasive technique for measuring the activities in the muscles. Instead of using a needle, an electrode is placed on the skin for measuring the myo-electric signals in the muscles. The EMG produces the output in the form of a voltage, which represents the amount of force a muscle is applying for certain activity.

With the progression of the remote framework and wearable identifying advancement, remote, wearable, and strong sensors confirm the affirmation of setting care in wireless body area network. The raw data are accumulated using wearable sensors from the person's body. After processing, the result is studied for diagnostics, health care, or security purposes [64]. One major purpose behind setting careful applications is the step by step activity watching, which screens a customer's natural signs, improvement precedents, or body present [64], and further works for development affirmation [65,66]. One major concern for wearable sensors is the client's consistency with wearing extra gadgets. The wearable sensors can be worn, or even inserted in garments, which will cause minimum obstacle for movement related applications [67,68]. It is found from the literature that gesture recognition, limb prosthetic control, gait analysis can be carried out with the help of EMG signals [69,70]. It provides the muscle activation pattern which make them able to identify the different activities [71].

3.4.2.1 MyoWare muscle sensor

MyoWare is a compact sEMG sensor which is developed by Advancer Technologies, shown in Fig. 3.6. It measures the activities in the muscle by taking into account the electric potential between the muscles. If brain commands for muscle flexion, it transfers an electric signal to the muscle, so that it starts gathering the motor units needed to carry out the action. The more the flexion, the more motor units are recruited, leading to increase in electrical activity of the muscle. This device analyses this myo-electric activity and generates analog signal as an output which demonstrates the flexion in the muscle.

The three holes on the right side of the MyoWare are the most essential. It simply provides the voltage to the sensor with + and - holes, and collects the (rectified and integrated) EMG signals from the sensor from "SIG." This output signal will be analog, ranging

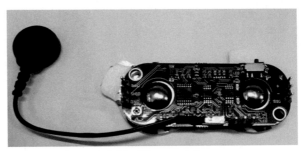

Figure 3.6 MyoWare muscle sensor.

in values 0–1023. To identify the RAW signal output, the sensor's top-leftmost side has a "RAW" signal output hole to collect the raw EMG signals. It can be visualized that "SHID"(+) and "GND"(−) are used to provide power to other electronic parts, if needed. "Gain" controls how much amount a user wants to amplify the EMG signal. Fig. 3.7 depicts the cable shield that is mounted on the sensor to use different set of electrodes.

A cost-effective wearable sensor-based clinical gait analysis provides a useful information in healthcare applications. The sEMG sensor-based gait analysis offers an important direction to analyze muscle pattern for detection of gait abnormality. A number of sEMG sensors are available in market for clinical applications. In our research, we are using MyoWare Muscle Sensor (AT-04-001) of Sparkfun cooperation which is very low-cost sensor with cost $40. The other available sEMG sensors are (A) Portable Open source Low-cost Electromyography (POLE) sensor, (B). Biometrics Ltd. sEMG sensors as wireless (LE230) or wired (SX230), and (C) Delsys sEMG sensor which comes with a high cost. The Sparkfun Myoware sensor is very lightweight and can be easily interfaced with Arduino Pro Mini board. It helps to transfer the data to the computer system. This sensor produces an amplified, rectified, and integrated output signal which provides an advantage in healthcare applications.

Figure 3.7 Cable shield.

3.4.3 Others: force sensitive resistors, goniometers

There exist other wearable sensors such as FSRs, goniometers which are often used in gait analysis. FSR sensors are used to extract the kinetic features such as GRF or plantar pressure. Multiple FSRs are attached with an insole to measure the applied force by different parts of the foot. The sensors are generally connected with an external microprocessor which stores the measured values. This complete setup with multiple FSRs is sometimes called as in-shoe pressure sensors or smart insoles. These sensors are often preferred over pressure mats or forceplates as the data can be captured for a long period of time without any space constraints. Tekscan F-scan systems, Kitronyx snowboard, etc., products accompany with software for users to visualize the captured signals. Thus, making these products a bit expensive and nonpreferable for cost-effective gait analysis. Fortunately, single FSR sensors are available in the market with price as low as \$20—\$50 from manufacturers such as Tekscan (Flexiforce), Sparkfun, etc. FSRs belonging to different price ranges (high-end, low-cost) are shown in Fig. 3.8.

Goniometers are used to measure angles, specifically joint angles in gait analysis. A traditional goniometer consists of a fixed arm, a moveable arm, and a fulcrum. To measure joint angle dynamically during any activities, electro-goniometers are used. There are mainly two types of electro-goniometer available: potentiometric and fiber-optic based. A potentiometric goniometer can be rigid or flexible depending on the structure.

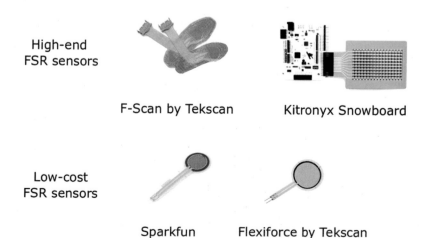

High-end
FSR sensors

F-Scan by Tekscan

Kitronyx Snowboard

Low-cost
FSR sensors

Sparkfun

Flexiforce by Tekscan

Figure 3.8 Force sensitive resistor sensors belonging to different categories.

In general, flexible potentiometric goniometers are used to capture angle dynamically for gait activities in literature. Goniometers manufactured by Biometrics limited have been frequently used in literature studies for its close to accurate measurements [72].

3.5 Summary

In this chapter, we discussed about different motion sensors that are generally used for gait analysis. We tried to focus more on the cost of different types of sensors and their setup. A detailed cost analysis is provided based on their purchase price available in the respective websites retrieved on September 2020. This chapter might be a helpful guide for users to select appropriate sensors in developing a cost-effective gait assessment system.

References

[1] S. Chakraborty, N. Thomas, A. Nandy, Gait abnormality detection in people with cerebral palsy using an uncertainty-based state-space model, Lect. Notes Comput. Sci. 12140 LNCS (Ml) (2020) 536–549, https://doi.org/10.1007/978-3-030-50423-6_40.

[2] P.E. Scranton JR, L.R. Pedegana, J.P. Whitesel, Gait analysis: alterations in support phase forces using supportive devices, Am. J. Sports Med. 10 (1) (1982) 6–11.

[3] I.A. Kramers-de Quervain, R. Müller, A. Stacoff, D. Grob, E. Stüssi, Gait analysis in patients with idiopathic scoliosis, Eur. Spine J. 13 (5) (2004) 449–456.

[4] H. Gerber, M. Zihlmann, M. Foresti, E. Stüssi, Method to simultaneously measure 3D kinematic and kinetic data during normal level walking using KISTLER force plates, VICON system and videofluoroscopy, J. Biomech. (2006).

[5] C.S. Nicolopoulos, E.G. Anderson, S.E. Solomonidis, P. V Giannoudis, Evaluation of the gait analysis FSCAN pressure system: clinical tool or toy? Foot 10 (3) (2000) 124–130.

[6] G. V Zammit, H.B. Menz, S.E. Munteanu, Reliability of the TekScan MatScan®system for the measurement of plantar forces and pressures during barefoot level walking in healthy adults, J. Foot Ankle Res. 3 (1) (2010) 11.

[7] A. Coda, T. Carline, D. Santos, Repeatability and reproducibility of the Tekscan HR-Walkway system in healthy children, Foot 24 (2) (2014) 49–55.

[8] F. Arafsha, C. Hanna, A. Aboualmagd, S. Fraser, A. El Saddik, Instrumented wireless smartinsole system for mobile gait analysis: a validation pilot study with tekscan strideway, J. Sens. Actuator Networks 7 (3) (2018) 36.

[9] V. Lugade, K. Kaufman, Dynamic stability margin using a marker based system and tekscan: a comparison of four gait conditions, Gait Posture 40 (1) (2014) 252–254.

[10] P. Catalfamo, D. Moser, S. Ghoussayni, D. Ewins, Detection of gait events using an F-scan in-shoe pressure measurement system, Gait Posture 28 (3) (2008) 420–426.

[11] K. Patrick, L. Donovan, Test–retest reliability of the Tekscan®F-Scan®7 in-shoe plantar pressure system during treadmill walking in healthy recreationally active individuals, Sport. Biomech. 17 (1) (2018) 83–97.

[12] H.L.P. Hurkmans, et al., Validity of the pedar mobile system for vertical force measurement during a seven-hour period, J. Biomech. 39 (1) (2006) 110–118.

[13] L.A. Boyd, E.L. Bontrager, S.J. Mulroy, J. Perry, The reliability and validity of the novel Pedar system of in-shoe pressure measurement during free ambulation, Gait Posture 2 (5) (1997) 165.

[14] A.K. Ramanathan, P. Kiran, G.P. Arnold, W. Wang, R.J. Abboud, Repeatability of the Pedar-X®in-shoe pressure measuring system, Foot Ankle Surg. 16 (2) (2010) 70–73.

[15] K. Kanitthika, K.S. Chan, Pressure sensor positions on insole used for walking analysis, in: The 18th IEEE International Symposium on Consumer Electronics (ISCE 2014), 2014, pp. 1–2.

[16] Q. Wang, G. Kurillo, F. Ofli, R. Bajcsy, Evaluation of pose tracking accuracy in the first and second generations of microsoft kinect, in: Proc. - 2015 IEEE Int. Conf. Healthc. Informatics, ICHI 2015, 2015, pp. 380–389, https://doi.org/10.1109/ICHI.2015.54.

[17] T. Guzsvinecz, V. Szucs, C. Sik-Lanyi, Suitability of the kinect sensor and leap motion controller-A literature review, Sensors 19 (5) (2019), https://doi.org/10.3390/s19051072.

[18] O. Wasenmüller, D. Stricker, Comparison of kinect v1 and v2 depth images in terms of accuracy and precision, Lect. Notes Comput. Sci. 10117, LNCS (2017) 34–45, https://doi.org/10.1007/978-3-319-54427-4_3.

[19] B. Müller, W. Ilg, M.A. Giese, N. Ludolph, Validation of enhanced kinect sensor based motion capturing for gait assessment, PLoS One 12 (4) (2017) 14–16, https://doi.org/10.1371/journal.pone.0175813.

[20] J. Shotton, et al., Real-time human pose recognition in parts from single depth images, Commun. ACM 56 (1) (2013) 116–124, https://doi.org/10.1145/2398356.2398381.

[21] N. García-Hernández, J. Corona-Cortés, L. García-Fuentes, R.D. González-Santibañez, V. Parra-Vega, Biomechanical and functional effects of shoulder kinesio taping® on cerebral palsy children interacting with virtual objects, Comput. Methods Biomech. Biomed. Eng. 22 (6) (2019) 676–684, https://doi.org/10.1080/10255842.2019.1580361.

[22] S. Chakraborty, A. Nandy, An unsupervised approach for gait phase detection, in: 4th Int. Conf. Comput. Intell. Networks, CINE 2020, 2020, pp. 1–5, https://doi.org/10.1109/CINE48825.2020.234396.

[23] E. Dolatabadi, B. Taati, A. Mihailidis, An automated classification of pathological gait using unobtrusive sensing technology, IEEE Trans. Neural Syst. Rehabil. Eng. 25 (12) (2017) 2336–2346, https://doi.org/10.1109/TNSRE.2017.2736939.

[24] M. Khokhlova, C. Migniot, A. Morozov, O. Sushkova, A. Dipanda, Normal and pathological gait classification LSTM model, Artif. Intell. Med. 94 (2019) 54–66, https://doi.org/10.1016/j.artmed.2018.12.007.

[25] S. Chakraborty, S. Jain, A. Nandy, G. Venture, Pathological gait detection based on multiple regression models using unobtrusive sensing technology, J. Signal Process. Syst. (2020), https://doi.org/10.1007/s11265-020-01534-1.

[26] S. Bei, Z. Zhen, Z. Xing, L. Taocheng, L. Qin, Movement disorder detection via adaptively fused gait analysis based on kinect sensors, IEEE Sens. J. 18 (17) (2018) 7305–7314, https://doi.org/10.1109/JSEN.2018.2839732.

[27] Q. Li, et al., Classification of gait anomalies from kinect, Vis. Comput. 34 (2) (2018) 229–241, https://doi.org/10.1007/s00371-016-1330-0.

[28] J. Latorre, C. Colomer, M. Alcañiz, R. Llorens, Gait analysis with the kinect v2: normative study with healthy individuals and comprehensive study of its sensitivity, validity, and reliability in individuals with stroke, J. Neuroeng. Rehabil. 16 (1) (2019) 1–11, https://doi.org/10.1186/s12984-019-0568-y.

[29] A.P. Rocha, H. Choupina, J.M. Fernandes, M.J. Rosas, R. Vaz, J.P.S. Cunha, "Kinect v2 based system for Parkinson's disease assessment, Proc. Annu. Int. Conf. IEEE Eng. Med. Biol. Soc. EMBS 2015 (November 2015) 1279–1282, https://doi.org/10.1109/EMBC.2015.7318601.

[30] S. Aahin, B. Köse, O.T. Aran, Z. Bahadlr Ağce, H. Kaylhan, The effects of virtual reality on motor functions and daily life activities in unilateral spastic cerebral palsy: a single-blind randomized controlled trial, Games Health J. 9 (1) (2020) 45–52, https://doi.org/10.1089/g4h.2019.0020.

[31] R. Cabrera, A. Molina, I. Gómez, J. García-Heras, Kinect as an access device for people with cerebral palsy: a preliminary study, Int. J. Hum. Comput. Stud. 108 (July 2017) 62–69, https://doi.org/10.1016/j.ijhcs.2017.07.004.

[32] W. Tao, T. Liu, R. Zheng, H. Feng, Gait analysis using wearable sensors, Sensors 12 (2) (2012) 2255–2283.

[33] V. Passaro, A. Cuccovillo, L. Vaiani, M. De Carlo, C.E. Campanella, Gyroscope technology and applications: a review in the industrial perspective, Sensors 17 (10) (2017) 2284.

[34] T. Cloete, C. Scheffer, Benchmarking of a full-body inertial motion capture system for clinical gait analysis, in: 2008 30th Annual International Conference of the IEEE Engineering in Medicine and Biology Society, 2008, pp. 4579–4582.

[35] J.-T. Zhang, A.C. Novak, B. Brouwer, Q. Li, Concurrent validation of Xsens MVN measurement of lower limb joint angular kinematics, Physiol. Meas. 34 (8) (2013) N63.

[36] T. Seel, J. Raisch, T. Schauer, IMU-based joint angle measurement for gait analysis, Sensors 14 (4) (2014) 6891–6909.

[37] M. Al-Amri, K. Nicholas, K. Button, V. Sparkes, L. Sheeran, J.L. Davies, Inertial measurement units for clinical movement analysis: reliability and concurrent validity, Sensors 18 (3) (2018) 719.

[38] J. Favre, B.M. Jolles, R. Aissaoui, K. Aminian, Ambulatory measurement of 3D knee joint angle, J. Biomech. 41 (5) (2008) 1029–1035.

[39] A V Dowling, J. Favre, T.P. Andriacchi, A wearable system to assess risk for anterior cruciate ligament injury during jump landing: measurements of temporal events, jump height, and sagittal plane kinematics, J. Biomech. Eng. 133 (7) (2011).

[40] H. Rouhani, J. Favre, X. Crevoisier, K. Aminian, Measurement of multi-segment foot joint angles during gait using a wearable system, J. Biomech. Eng. 134 (6) (2012).

[41] S. Bakhshi, M.H. Mahoor, B.S. Davidson, Development of a body joint angle measurement system using IMU sensors, in: 2011 Annual International Conference of the IEEE Engineering in Medicine and Biology Society, 2011, pp. 6923–6926.

[42] T. Jaysrichai, A. Suputtitada, W. Khovidhungij, Mobile sensor application for kinematic detection of the knees, Ann. Rehabil. Med. 39 (4) (2015) 599.

[43] Z. Yang, et al., Validation of a novel gait analysis system, in: 2017 IEEE/ACM International Conference on Connected Health: Applications, Systems and Engineering Technologies (CHASE), 2017, pp. 288–289.

[44] W.W. Lee, et al., A smartphone-centric system for the range of motion assessment in stroke patients, IEEE J. Biomed. Heal. informatics 18 (6) (2014) 1839–1847.

[45] Y. Kumar, et al., Wireless wearable range-of-motion sensor system for upper and lower extremity joints: a validation study, Healthc. Technol. Lett. 2 (1) (2015) 12–17.

[46] T. Nakano, et al., Gaits classification of normal vs. patients by wireless gait sensor and Support Vector Machine (SVM) classifier, Int. J. Softw. Innov. 5 (1) (2017) 17–29.

[47] N.U. Ahamed, D. Kobsar, L.C. Benson, C.A. Clermont, S.T. Osis, R. Ferber, Subject-specific and group-based running pattern classification using a single wearable sensor, J. Biomech. 84 (2019) 227–233.

[48] N.U. Ahamed, D. Kobsar, L. Benson, C. Clermont, R. Kohrs, S.T. Osis, R. Ferber, Using wearable sensors to classify subject-specific running biomechanical gait patterns based on changes in environmental weather conditions, PLoS One 13 (2018) e0203839.

[49] H.J. Luinge, P.H. Veltink, C.T.M. Baten, "Estimation of orientation with gyroscopes and accelerometers, in: Proceedings of the First Joint BMES/EMBS Conference, IEEE Engineering in Medicine and Biology 21st Annual Conference and the 1999 Annual Fall Meeting of the Biomedical Engineering Society (Cat. N, 1999), vol. 2, 1999, p. 844.

[50] G. Cooper, et al., Inertial sensor-based knee flexion/extension angle estimation, J. Biomech. 42 (16) (2009) 2678–2685.

[51] H. Saito, T. Watanabe, Kalman-filtering-based joint angle measurement with wireless wearable sensor system for simplified gait analysis, IEICE Trans. Inf. Syst. 94 (8) (2011) 1716–1720.

[52] A.D. Young, Comparison of orientation filter algorithms for realtime wireless inertial posture tracking, in: 2009 Sixth International Workshop on Wearable and Implantable Body Sensor Networks, 2009, pp. 59–64.

[53] A. Mannini, A.M. Sabatini, A hidden Markov model-based technique for gait segmentation using a foot-mounted gyroscope, in: 2011 Annual International Conference of the IEEE Engineering in Medicine and Biology Society, 2011, pp. 4369–4373.

[54] R.L. Evans, D.K. Arvind, Detection of gait phases using orient specks for mobile clinical gait analysis, in: 2014 11th International Conference on Wearable and Implantable Body Sensor Networks, 2014, pp. 149–154.

[55] I.H. López-Nava, A. Muñoz-Meléndez, A.I. Pérez Sanpablo, A. Alessi Montero, I. Quiñones Urióstegui, L. Núñez Carrera, Estimation of temporal gait parameters using Bayesian models on acceleration signals, Comput. Methods Biomech. Biomed. Eng. 19 (4) (2016) 396–403.

[56] W.-C. Hsu, et al., Multiple-wearable-sensor-based gait classification and analysis in patients with neurological disorders, Sensors 18 (10) (2018) 3397.

[57] A. Mannini, D. Trojaniello, A. Cereatti, A.M. Sabatini, A machine learning framework for gait classification using inertial sensors: application to elderly, post-stroke and huntington's disease patients, Sensors 16 (1) (2016) 134.

[58] A. Elkurdi, I. Caliskanelli, S. Nefti-Meziani, Amplitude modulation and convolutional encoder techinques for gait speed classification, in: 2018 23rd International Conference on Methods & Models in Automation & Robotics (MMAR), 2018, pp. 544–549.

[59] Y. Xia, J. Zhang, Q. Ye, N. Cheng, Y. Lu, D. Zhang, Evaluation of deep convolutional neural networks for detection of freezing of gait in

Parkinson's disease patients, Biomed. Signal Process. Control 46 (2018) 221–230.

[60] D. Ravi, C. Wong, B. Lo, G.-Z. Yang, A deep learning approach to on-node sensor data analytics for mobile or wearable devices, IEEE J. Biomed. Heal. Inf. 21 (1) (2016) 56–64.

[61] J. Camps, et al., Deep learning for freezing of gait detection in Parkinson's disease patients in their homes using a waist-worn inertial measurement unit, Knowledge-Based Syst. 139 (2018) 119–131.

[62] J. Chakraborty, A. Nandy, Discrete wavelet transform based data representation in deep neural network for gait abnormality detection, Biomed. Signal Process. Control 62 (2020) 102076.

[63] O. Dehzangi, M. Taherisadr, R. ChangalVala, IMU-based gait recognition using convolutional neural networks and multi-sensor fusion, Sensors 17 (12) (2017) 2735.

[64] D.P. Tobón, T.H. Falk, M. Maier, Context awareness in WBANs: a survey on medical and non-medical applications, IEEE Wirel. Commun. 20 (4) (2013) 30–37.

[65] S.H. Roy, et al., A combined sEMG and accelerometer system for monitoring functional activity in stroke, IEEE Trans. Neural Syst. Rehabil. Eng. 17 (6) (2009) 585–594.

[66] J. Cheng, X. Chen, M. Shen, A framework for daily activity monitoring and fall detection based on surface electromyography and accelerometer signals, IEEE J. Biomed. Heal. Inf. 17 (1) (2012) 38–45.

[67] S. Thongpanja, A. Phinyomark, P. Phukpattaranont, C. Limsakul, Mean and median frequency of EMG signal to determine muscle force based on time-dependent power spectrum, Elektron. Elektrotechnika 19 (3) (2013) 51–56.

[68] C. Zhu, W. Sheng, Wearable sensor-based hand gesture and daily activity recognition for robot-assisted living, IEEE Trans. Syst. Man, Cybern. A Syst. Humans 41 (3) (2011) 569–573.

[69] A.A. Adewuyi, L.J. Hargrove, T.A. Kuiken, An analysis of intrinsic and extrinsic hand muscle EMG for improved pattern recognition control, IEEE Trans. Neural Syst. Rehabil. Eng. 24 (4) (2015) 485–494.

[70] I. Vujaklija, D. Farina, O.C. Aszmann, New developments in prosthetic arm systems, Orthop. Res. Rev. 8 (2016) 31.

[71] M.-S.S. Cheng, Monitoring Functional Motor Activities in Patients with Stroke, Boston University, 2005.

[72] G. Campiglio, J. Mazzeo, S. Rodriguez, A wireless goniometry system, J. Phys. Conf. Series 7 (2013) 12008.

4

Validation study of low-cost sensors

4.1 Introduction

Assessment of kinematic variables or features, such as joint angles of pelvis, hip, knee, ankle, etc., is extremely important in clinical gait analysis. These parameters are used in different purposes like assessment of abnormal gait [1], risk of fall assessment in geriatric population [2], etc. Traditionally, gait analysis, these features are computed from the time series data acquired from the gold standard systems. But, these sensors are highly expensive and unaffordable for many clinics. As an alternative approach, some low-cost sensors like Kinect, IMU, etc., have been used for various clinical applications [3–5] and gradually have become very popular. Validation testing is an important prerequisite for using such sensors in clinical purpose.

4.2 Kinect validation for clinical usages

In vision-based gait analysis, Kinect has become very popular as a cost-effective sensor. Mainly, its skeletal data stream has been used extensively for clinical purpose. Hence, different studies have tried to validate the skeletal data stream for gait features. Table 4.1 demonstrates some state-of-the-art methods for Kinect validation. It can be seen that researches have been conducted both on over-ground and treadmill.

Treadmill validation studies have been performed primarily focusing on normal gait. Pfister et al. [6] tested the validity of skeleton data stream of Kinect using a treadmill experiment (see Fig. 4.1 (I)). Subjects were asked to walk on three different predefined speeds. Data were captured simultaneously from 10 Vicon cameras and a Kinect placed at 45° with respect to the treadmill. Maximum flexion and extension angle of hip and knee and stride timing were computed for both systems. They reported a good agreement for stride time. For kinematic feature, although Kinect

Modern Methods for Affordable Clinical Gait Analysis. https://doi.org/10.1016/B978-0-323-85245-6.00005-9
Copyright © 2021 Elsevier Inc. All rights reserved.

Table 4.1 State-of-the-art methods for Kinect validation.

Article	Populations	Experiment type	Ground truth system	Kinect type (number)	Data stream	Feature validated	Findings
Pfister et al. [6]	Normal	Treadmill	Vicon	Kinect v1 (1)	Skeletal	max. flexion and extension angles of hip and knee, stride time	Good agreement for stride time, but, for joint kinematics, moderate error was observed (RMSE > 5°)
Xu et al. [7]	Normal	Treadmill	Vicon	Kinect v2 (1)	Skeletal	Spatio-temporal (SPT), joint kinematics	Fair agreement for SPT, but, for joint kinematics, error was high (RMSE = 20.15°)
Eltoukhy et al. [8]	Normal	Treadmill	Vicon	Kinect v1 and v2 (1)	Skeletal	SPT, sagittal plane hip, knee, and ankle ROM	Good agreement for SPT, hip, and knee range of motion (ROM), for ankle ROM, error was High
Macpherson et al. [9]	Normal	Treadmill	Qualisys	Kinect v1 (1)	Skeletal	Linear and angular kinematics for trunk and pelvic joints	High agreement for linear kinematics, but, high error for angular kinematics
Ma et al. [10]	Normal, cerebral palsy	Overground	Vicon	Kinect v2 (1)	Skeletal	Joint kinematics	Agreement for joint kinematics was not clinically acceptable
Boas et al. [11]	Normal	Overground	Vicon	Kinect v2 (1)	Skeletal	Linear and angular kinematics	For linear kinematics, between system agreement was good, but, for angular kinematics, error was high
Tanaka et al. [12]	Normal	Overground	Qualisys	Kinect v2 (1)	Skeletal	Sagittal and frontal hip and knee joint angles	Except sagittal plane knee angle, between system agreement was not clinically acceptable.
Stone et al. [13]	Normal	Overground	Qualisys	Kinect v2 (1)	Raw depth image	SPT and kinematics	Good agreement obtained for SPT, but, for kinematics, error was high
Geerse et al. [14]	Normal	Overground	Vicon	Kinect v2 (4)	Skeletal	SPT	Good agreement obtained
Muller et al. [15]	Normal	Overground	Vicon	Kinect v2 (6)	Skeletal	SPT	Good agreement obtained

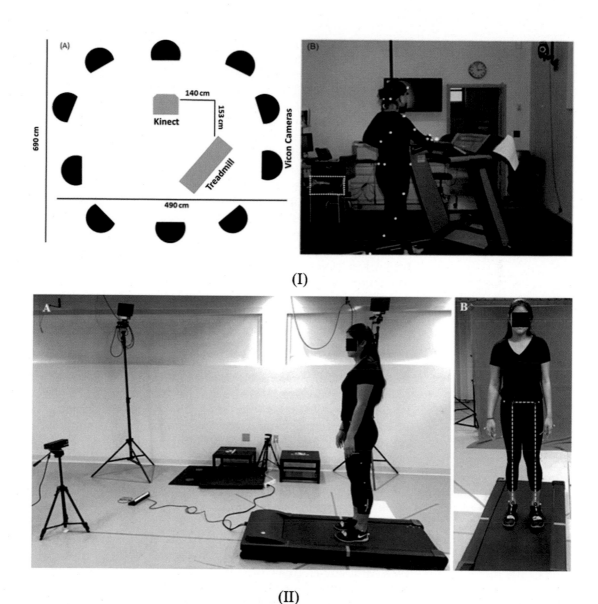

(I)

(II)

Figure 4.1 Different setups for Kinect validation. (I) Treadmill and camera layout. (A) The schematic displays the equipment arrangement. (B) The photograph of the camera and treadmill layout shows the Kinect sensor highlighted by the white dashed rectangle to the left of the subject. (II) (A) The experimental setup including eight BTS infrared cameras, one Kinect v2, and the Life Span treadmill. (B) Reflective markers attached to the subject's lower extremity and the Kinect skeleton (hip, knee, and ankle joints).

followed the pattern of movement trajectory of the ground truth, the magnitude of error was high (>5°).

Xu et al. [7] performed a slightly different experiment with same objective. Here the Kinect was placed in front of the treadmill. They also reported a fair agreement for various spatio-temporal (SPT) features, but for joint kinematics, results were not up to a satisfactory level (average RMSE 20.15°). Eltoukhy et al. [8] also placed the Kinect in front of the treadmill and performed the validation experiment (see Fig. 4.1 (II)). They obtained an acceptable between systems agreement for SPT features and sagittal plane knee and hip joint range of motion (ROM) and joint angle, but, for ankle joint kinematics, the results were not impressive. However, the agreement on magnitude of joint angle of knee and hip varied across different phases of gait cycle. Ma et al. [10] tested Kinect for the population of cerebral palsy (CP) patients. Validation was performed during over-ground walking of subjects. A calibration method based on linear regression and long short-term memory was proposed to increase the between systems agreement. The authors reported Kinect as an unacceptable sensor for measuring joint kinematics (coefficients of multiple correlation = 0.001 to 0.7). Boas et al. [11] performed validity testing for both versions of Kinect. Subjects were asked to walk toward and away from the Kinect. The authors obtained a promising result for linear kinematics for Kinect v2 while subjects were walking toward the camera. However, for angular kinematics, between systems agreement was not up to a satisfactory level. They concluded that the both sensors can be used as an alternative of the expensive gold standard system, especially to measure some linear and angular parameters (namely knee joint angle). Tanaka et al. [12] validated the hip and knee joint angles (sagittal and frontal planes) obtained from skeleton data stream of Kinect v2. Parallel translation of the time series data of Kinect was performed to align (approximately) both the system data. The authors observed similar trajectories in sagittal plane for both systems, while in frontal plane, the pattern was different from the reference system. However, except the sagittal plane knee angle, error in joint kinematics between the systems is < 5° which is considered as marginal acceptable error for clinical validation [16].

Stone et al. [13] constructed a two-Kinect setup and validated some SPT features. Data acquisition from each Kinect performed independently. Instead of skeletal image, the authors used three-dimensional point cloud to compute the SPT features. For the temporal features, a good agreement was reported. A few studies have constructed a multi-Kinect setup (mainly on the overground) and validated it against the gold standard. Geerse et al. [14] a four-Kinect v2 architecture which cover 10m walking path. Kinects were placed at one side of the walking path. Coordinates for both of the systems (i.e., Kinects and the gold standard system) were aligned using a spatial calibration grid. Subjects were asked to walk both at self-selected and maximum possible speeds. In the preprocessing step, each Kinect body-point data were treated separately. If in a frame at least 15 joints were not identified, then it was treated as invalid. If more than one Kinect captures a valid frame of movement, then mean value for the associated 3D positions was computed. For both body-point time series and SPT features between-systems, agreement was high (except step width). Muller et al. [15] modified the architecture proposed by Geerse et al. [14] by placing Kinects at both sides of the walking track. A client-server protocol was established to control the six Kinects system. All the computers were synchronized using precision time protocol. Different software modules were implemented in the server to manage, record, and calibrate multiple sensors outputs. Point cloud from each client was collected by the server and transformed to the global coordinate system using spatial calibration which generated a combined point cloud of body surface and skeleton. The authors obtained an excellent between systems agreement [Pearson's correlation coefficient (PCC) (average) > 0.9] for the SPT features.

It is noteworthy that almost all of the above-mentioned studies (both treadmill and overground) have reported Kinect as a suitable sensor to measure SPT features, but for kinematic features, the sensor was not recommended. This is mainly because of the limited FoV of a single Kinect system which also arises from the partial self-occlusion problem. Again, all the experiments conducted on the skeletal data stream. Since, obtaining the body point time series data is comparatively easier from the skeletal data stream, it has been the primary attraction point for the clinical gait analysis. Studies have often avoided the potential of the high-resolution color image data stream to provide the kinematic parameters. Recently, Chakraborty et al. [17] have proposed a methodology to extract joint angle from color image data stream of Kinect v2. The experiment was conducted on treadmill on normal subjects. Point cloud feature of Kinect was used to extract lower limb joint position, which was then converted to joint angle using extended Kalman filter and a kinematic model. The process was validated against the gold standard cameras. The obtained joint angles were significant to use for medical purposes.

4.3 Inertial sensor validation on estimating joint angles

To validate the usage of inertial sensors in medical domain, it is important to validate and verify the preciseness of kinematic features with respect to standard MoCap systems. Correct estimation of joint angle helps in measuring the kinematic features correctly. Accelerometer is useful to measure angular displacement, however, it is prone to drift and noise [18,19]. Accelerometer provides information about static orientation with respect to gravity, whereas gyroscope provides information about angular velocity with respect to the sensor's axis which is useful for highly dynamic movements (e.g., gait activities). On the other hand, accelerometer measurements fused with magnetometer measurements are useful in estimating slow-motion displacements. The general method of obtaining joint angle is by integrating angular velocity over a period of time. However, when placed on the body, the data get biased because of drift. These can be rectified using information on azimuth using magnetometer which is more commonly used in popular MEMS sensors (Xsens). Recently, many recent research studies propose new approaches of estimating joint angle without using magnetometer but combining values of gyroscope and magnetometer [20,21]. In this approach, first joint angle is measured by integrating gyroscope values, then corrected by estimation based on inclination data from accelerometers collected during periods of rest, or near constant velocity motion with the help of filters such as Kalman filter [20] or complimentary filter [22]. The joint angle can be accurately estimated if the joint axis position with respect to local coordinate systems and the orientation of the sensors are known [23]. The complete orientation of IMU sensors with respect to a global reference coordinate system can be calculated by fusing gyroscope and accelerometer measurements. Researchers have extensively studied about the validity of IMU sensors which combines accelerometer, gyroscope, and magnetometer to estimate the joint angle during various gait activities and performing tasks. Among different lower-limb angles, majority of the studies estimate the knee flexion/extension angle [21,24,25]. Some of the studies estimated joint angles for hip, knee, and ankle [26–28]. The knee flexion/extension angle can be derived when IMU sensors are placed around the upper leg (thigh) and lower leg (shank) as shown in Fig. 4.2A. The sensors can be placed on the limbs with different orientations. However, Palermo et al. reported that position of the sensors has negligible effect on the angle estimation [26]. Similarly, hip angle can be estimated using the values obtained from trunk and shank as shown in Fig. 4.2B; ankle angle can be estimated from shank and

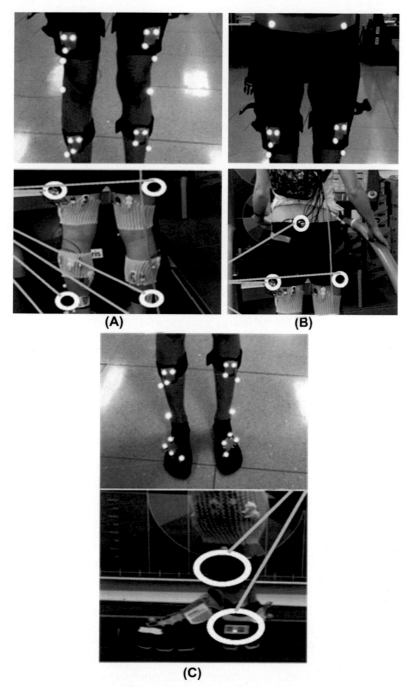

Figure 4.2 Placement of IMU sensors for estimating lower-limb joint angles (A) knee (B) hip (C) ankle [26,27].

foot as shown in Fig. 4.2C. Anatomical constraints on joints can be helpful during angle estimation [21]. There have been three significant review articles which discussed about the validity and reliability of joint angle estimation methods using wearable IMU sensors [29–31]. Authors in Ref. [31] suggested that for lower limb joints, IMUs could be a good alternative to motion capture systems for movement measurements in the sagittal and frontal planes, but not in the transverse plan.

4.3.1 Evaluation metrics for validation of estimated joint angles

Validity of IMU sensor is performed by comparing between the angles obtained by IMU sensors and standard sensors. The evaluation metrics used for validation are as follows:

(1) Correlation coefficients—Correlation coefficient signifies the relation between two signals. Its value less than 0.5 is considered poor, whereas between 0.5 and 0.75 is considered moderate correlation between signals, between 0.75 and 0.90 is considered good and greater than 0.90 is excellent [32]. There are multiple variations of correlation coefficients.

  (a) Sample correlation coefficient—Sample correlation coefficient is the ratio between sample covariance of two variable and product of the sample standard deviation of each variable.

  (b) PCC—It is the ratio between population covariance and product of population standard deviation values of two variables.

  (c) Coefficient of determination—This is determined by squaring the correlation coefficient which shows how strongly a variability in one signal effect the variability of another signal.

  (d) Coefficient of multiple correlation (CMC)—CMC is measured by PCC of actual values of a variable and the predicted or estimated values of the variable using (multiple) independent variables.

  (e) Intraclass correlation coefficient (ICC)—This signifies how strongly the signal is related with itself. The modern ICC is a modified version of PCC. The estimators used for modern ICC can be defined as random effects model, $Y_{ij} = \mu + \alpha_j + \eta_{ij}$ where Y is the *ith* observation of variable j and μ is mean, α_j is unobserved random effect and η_{ij} is unobserved random noise. The population ICC can be derived as the following: $\frac{\sigma_\alpha^2}{\sigma_{\alpha^2} + \sigma_\eta^2}$.

Table 4.2 Ranges of error metric values obtained by different studies as mentioned in [31].

Joint angle	RMSE	Correlation coefficient
Hip	0.2–9.3°	0.53 to 1.00
Knee	1–11.5°	0.4 to 1.00
Ankle	0.4–18.8°	0.33 to 0.99

(2) Root mean square error (RMSE/RMS)—It is the measurement error between the reference system and the IMU sensors. It is measured by squaring the difference of the values obtained from both sensors. Error (RMSE/RMS) less than 5° is considered as excellent and between 5 and 10° as good.

(3) Limits of agreement—It is used to present the estimated range/interval of the difference of two measurements lie. Here the two measurements are joint angles obtained from the standard device and joint angles obtained from IMU sensors. Table 4.2 discusses the ranges of the evaluation metrics obtained by various studies for different joint positions in the lower limb using IMU sensors.

Analyzing the findings of many literature studies, it can be stated that IMU sensors can be beneficial for healthcare systems for measuring kinematic features up to certain extent of precision. There are some studies which fused IMU data with Kinect data for joint angle estimation which showed promising results and can be explored further [33,34]. However, most of the studies considered healthy subjects for validation of the sensor and limited to general tasks such as walking, running, sitting, etc. Researchers should be encouraged to perform the validation for subjects with gait disorders, for performing both simple and complex tasks such as climbing stairs, stepping over hurdles, etc.

4.4 Summary

This chapter addresses an important aspect of effectiveness of affordable sensors in clinical practice, i.e., the validity of the measured kinematics features against standard sensors. Different studies which performed the validity tests of Kinect and IMU sensors are presented. The methodologies used and reports of these studies are discussed. Positive reports with small range of error indicates that there is high possibility of using these sensors in clinical assessment in near future.

References

[1] A. Malone, D. Kiernan, H. French, V. Saunders, T. O'Brien, Do children with cerebral palsy change their gait when walking over uneven ground? Gait Posture 41 (2) (2015) 716–721, https://doi.org/10.1016/j.gaitpost.2015.02.001.

[2] G. Kemoun, P. Thoumie, D. Boisson, J.D. Guieu, Ankle dorsiflexion delay can predict falls in the elderly, J. Rehabil. Med. 34 (6) (2002) 278–283.

[3] E. Dolatabadi, B. Taati, A. Mihailidis, An automated classification of pathological gait using unobtrusive sensing technology, IEEE Trans. Neural Syst. Rehabil. Eng. 25 (12) (2017) 2336–2346, https://doi.org/10.1109/TNSRE.2017.2736939.

[4] W. Zhao, H. Feng, R. Lun, D.D. Espy, M.A. Reinthal, A Kinect-Based Rehabilitation Exercise Monitoring and Guidance System, 2014, https://doi.org/10.1109/ICSESS.2014.6933678.

[5] S. Bei, Z. Zhen, Z. Xing, L. Taocheng, L. Qin, Movement disorder detection via adaptively fused gait analysis based on kinect sensors, IEEE Sensor. J. 18 (17) (2018) 7305–7314, https://doi.org/10.1109/JSEN.2018.2839732.

[6] A. Pfister, A.M. West, S. Bronner, J.A. Noah, Comparative abilities of Microsoft Kinect and Vicon 3D motion capture for gait analysis, J. Med. Eng. Technol. 38 (5) (2014) 274–280.

[7] X. Xu, R.W. McGorry, L.S. Chou, J. Hua Lin, C. Chi Chang, Accuracy of the Microsoft Kinect^{TM} for measuring gait parameters during treadmill walking, Gait Posture 42 (2) (2015) 145–151, https://doi.org/10.1016/j.gaitpost.2015.05.002.

[8] M. Eltoukhy, J. Oh, C. Kuenze, J. Signorile, Improved kinect-based spatiotemporal and kinematic treadmill gait assessment, Gait Posture (2017), https://doi.org/10.1016/j.gaitpost.2016.10.001.

[9] T.W. Macpherson, J. Taylor, T. McBain, M. Weston, I.R. Spears, Real-time measurement of pelvis and trunk kinematics during treadmill locomotion using a low-cost depth-sensing camera: a concurrent validity study, J. Biomech. 49 (3) (2016) 474–478, https://doi.org/10.1016/j.jbiomech.2015.12.008.

[10] Y. Ma, K. Mithraratne, N.C. Wilson, X. Wang, Y. Ma, Y. Zhang, The validity and reliability of a Kinect v2-based gait analysis system for children with cerebral palsy, Sensors 19 (7) (2019) 1660.

[11] M. do C. Vilas-Boas, H.M.P. Choupina, A.P. Rocha, J.M. Fernandes, J.P.S. Cunha, Full-body motion assessment: concurrent validation of two body tracking depth sensors versus a gold standard system during gait, J. Biomech. 87 (2019) 189–196, https://doi.org/10.1016/j.jbiomech.2019.03.008.

[12] R. Tanaka, H. Takimoto, T. Yamasaki, A. Higashi, Validity of time series kinematical data as measured by a markerless motion capture system on a flatland for gait assessment, J. Biomech. 71 (2018) 281–285, https://doi.org/10.1016/j.jbiomech.2018.01.035.

[13] E.E. Stone, M. Skubic, Passive in-home measurement of stride-to-stride gait variability comparing vision and Kinect sensing, in: 2011 Annual International Conference of the IEEE Engineering in Medicine and Biology Society, 2011, pp. 6491–6494.

[14] D.J. Geerse, B.H. Coolen, M. Roerdink, Kinematic validation of a multi-Kinect v2 instrumented 10-meter walkway for quantitative gait assessments, PLoS One 10 (10) (2015) 1–15, https://doi.org/10.1371/journal.pone.0139913.

[15] B. Müller, W. Ilg, M.A. Giese, N. Ludolph, Validation of enhanced kinect sensor based motion capturing for gait assessment, PLoS One 12 (4) (2017) 14–16, https://doi.org/10.1371/journal.pone.0175813.

[16] J.L. McGinley, R. Baker, R. Wolfe, M.E. Morris, The reliability of three-dimensional kinematic gait measurements: a systematic review, Gait Posture 29 (3) (2009) 360–369.

[17] S. Chakraborty, A. Nandy, T. Yamaguchi, V. Bonnet, G. Venture, Accuracy of image data stream of a markerless motion capture system in determining the local dynamic stability and joint kinematics of human gait, J. Biomech. (2020) 109718.

[18] G. Grimaldi, M. Manto, Neurological tremor: sensors, signal processing and emerging applications, Sensors 10 (2) (2010) 1399–1422.

[19] H. Dejnabadi, B.M. Jolles, K. Aminian, A new approach to accurate measurement of uniaxial joint angles based on a combination of accelerometers and gyroscopes, IEEE Trans. Biomed. Eng. 52 (8) (2005) 1478–1484.

[20] J. Favre, B.M. Jolles, R. Aissaoui, K. Aminian, Ambulatory measurement of 3D knee joint angle, J. Biomech. 41 (5) (2008) 1029–1035.

[21] G. Cooper, et al., Inertial sensor-based knee flexion/extension angle estimation, J. Biomech. 42 (16) (2009) 2678–2685.

[22] A.D. Young, Comparison of orientation filter algorithms for realtime wireless inertial posture tracking, in: 2009 Sixth International Workshop on Wearable and Implantable Body Sensor Networks, 2009, pp. 59–64.

[23] T. Seel, J. Raisch, T. Schauer, IMU-based joint angle measurement for gait analysis, Sensors 14 (4) (2014) 6891–6909.

[24] T. McGrath, R. Fineman, L. Stirling, An auto-calibrating knee flexion-extension axis estimator using principal component analysis with inertial sensors, Sensors 18 (6) (2018) 1882.

[25] J. Favre, R. Aissaoui, B.M. Jolles, J.A. de Guise, K. Aminian, Functional calibration procedure for 3D knee joint angle description using inertial sensors, J. Biomech. 42 (14) (2009) 2330–2335.

[26] E. Palermo, S. Rossi, F. Marini, F. Patanè, P. Cappa, Experimental evaluation of accuracy and repeatability of a novel body-to-sensor calibration procedure for inertial sensor-based gait analysis, Measurement 52 (2014) 145–155.

[27] C. Nüesch, E. Roos, G. Pagenstert, A. Mündermann, Measuring joint kinematics of treadmill walking and running: comparison between an inertial sensor based system and a camera-based system, J. Biomech. 57 (2017) 32–38.

[28] E. Dorschky, M. Nitschke, A.-K. Seifer, A.J. van den Bogert, B.M. Eskofier, Estimation of gait kinematics and kinetics from inertial sensor data using optimal control of musculoskeletal models, J. Biomech. 95 (2019) 109278.

[29] A.I. Cuesta-Vargas, A. Galán-Mercant, J.M. Williams, The use of inertial sensors system for human motion analysis, Phys. Ther. Rev. 15 (6) (2010) 462–473.

[30] C.P. Walmsley, S.A. Williams, T. Grisbrook, C. Elliott, C. Imms, A. Campbell, Measurement of upper limb range of motion using wearable sensors: a systematic review, Sport. Med. 4 (1) (2018) 53.

[31] I. Poitras, et al., Validity and reliability of wearable sensors for joint angle estimation: a systematic review, Sensors 19 (7) (2019) 1555.

[32] I. Pasciuto, G. Ligorio, E. Bergamini, G. Vannozzi, A.M. Sabatini, A. Cappozzo, How angular velocity features and different gyroscope noise types interact and determine orientation estimation accuracy, Sensors 15 (9) (2015) 23983–24001.

[33] A.P.L. Bo, M. Hayashibe, P. Poignet, Joint angle estimation in rehabilitation with inertial sensors and its integration with Kinect, in: 2011 Annual International Conference of the IEEE Engineering in Medicine and Biology Society, 2011, pp. 3479–3483.

[34] A. Akbari, X. Thomas, R. Jafari, Automatic noise estimation and context-enhanced data fusion of IMU and kinect for human motion measurement, in: 2017 IEEE 14th International Conference on Wearable and Implantable Body Sensor Networks (BSN), 2017, pp. 178–182.

5

Gait segmentation and event detection techniques

5.1 Introduction

The transition point from one sub-phase to another in a gait cycle is referred as gait event. Annotation of gait events such as heel strike (HS), heel-off (HO), flat foot (FF), toe-off (TO), etc., is vital to process and analyze the gait pattern. It is considered as the initial step for gait analysis. Gait phases, i.e., time duration between two consecutive events, are also important to extract vital clinical information. Generally, in clinic manual annotation technique is followed. But, it is laborious, time consuming, and error prone. Instead, automated techniques have become a popular alternative. On this direction, simple threshold-based methods or machine learning algorithms have been used for event detection. Studies have used different sensors for acquiring data which were used to partition gait cycle [1—8]. Generally, force plate data or gold standard cameras or manual annotation procedure is used to construct the ground truth for event detection. Researches have been conducted to segment the gait cycle in two or more granularity levels. However, the proper choice of granularity level depends on the sensor, specific application, and computational method [9].

5.2 Why gait cycle segmentation?

Extraction of different spatio-temporal features highly depends on the proper annotation of gait events. Cycle segmentation is crucial for different clinical applications also. For example, while designing functional electrical stimulation, prostheses, exoskeletons, orthoses, etc., partitioning of gait cycle act as a control variable [10]. Sub-phase wise analysis of gait is possible only after segmenting the gait cycle.

Modern Methods for Affordable Clinical Gait Analysis. https://doi.org/10.1016/B978-0-323-85245-6.00002-3

5.3 Vision sensor-based gait cycle segmentation

Different studies have used vision-based sensors to segment gait cycle. Table 5.1 represents some popular state-of-the-art which can be broadly divided into threshold-based and machine learning–based methods.

5.3.1 Threshold-based methods

Zeni et al. [11] proposed two algorithms for treadmill and overground gait to detect HS and TO events. Positional data from the heel and toe markers were collected using the gold standard cameras. GRF data obtained from force plate were used as the ground truth. Both normal and pathological subjects (multiple sclerosis and stroke) participated in this experiment. In case of the coordinate-based algorithm for treadmill, foot markers data [anterior posterior (AP) direction] were subtracted from the corresponding sacral data to obtain a sinusoidal curve. Mean value of this curve was selected as the threshold. The peak values beyond this threshold were marked as the HS points whereas the valleys were TO points. In case of velocity-based algorithm for treadmill, changes of sign of the velocity vector (AP direction) of the markers were used to identify the events. For overground gait, each of the marker data of body (for each frame) were subtracted from the corresponding sacral data and the same two algorithms were applied after that. The authors obtained a competing result. O'Connor et al. [12] used vertical velocity of foot to detect HS and TO events during an overground experiment. Heel and toe marker data obtained from the gold standard cameras were used to create the foot center. Local maximum of vertical velocity of foot center within a predefined window was marked as the TO point. A comparatively smaller size window was used to identify a set of troughs which were marked as HS points. Then a threshold was build based on a heuristic to select the correct HS point. The algorithm was applied on both normal and pathological (spastic diplegia) populations and validated against the GRF data from force plate. Tang et al. [13] used a two-dimensional camera to calculate the TO events. The authors have used a publicly available CASIA gait database [19]. Different sizes of consecutive silhouettes difference (CSD) maps were generated by combining consecutive silhouette images of a pedestrian. After that each CSD was normalized. Convolutional neural network was used to detect TO events from the CSD-maps. Auvinet et al. [2] computed

Table 5.1 State-of-the-art methods for gait event detection using vision-based sensors.

Article	Populations	Data acquisition	Experiment type	Ground truth	Feature	Method used	Results
Zeni et al. [11]	Stroke, MS, Normal	Vicon	Overground Treadmill	Force plate	Foot joint positions and velocity	Threshold-based algorithm	Two frame error
Connor et al. [12]	Spastic diplegia, Normal	Vicon	Overground	Force plate	Vertical velocity of heel and toe	Threshold-based algorithm	Three frame error
Tang et al. [13]	CASIA gait database	Two-dimensional camera	N.A	CASIA gait database	Consecutive silhouettes Difference map	Convolutional neural network	Accuracy: 82%
Auvinet et al. [2]	Normal	Kinect v1	Treadmill	Vicon	Knee distance	Threshold-based algorithm	Four frame error
Lambrecht et al. [14]	Normal	Vicon	Treadmill	Force plate	Kinematic data	Threshold-based algorithm	Three frame error
Miller et al. [15]	CP, normal	Vicon	Overground	Force plate	Foot kinematics	Neural network	Two frame error
Kidzinsk et al. [16]	CP, normal	Qualisys	Overground	Manual annotation	Lower limb joint positions and velocity	Long short-term memory (LSTM)	Two frame error
Tanghe et al. [17]	Normal	Qualisys	Treadmill	Manual annotation	Lower limb joint velocity	Bayesian probabilistic model	Two frame error
Lempereur et al. [18]	CP, stroke, MS, normal	Vicon	Overground	Force plate	Foot marker position and velocity	Bidirectional LSTM	Two frame error

HS events during treadmill walking of normal healthy subjects. Data were acquired from a Kinect sensor. Ground truth was generated from the gold standard cameras. The authors hypothesized that HS will occur when the distance between knees (longitudinal axis) would be maximum. Knee distance was computed in three different methods. In the first method, the gold standard system was used to estimate the distance using knee joints positions. In the second method, knee joints were localized on the depth map of Kinect using anthropometric data. Third method used the same anthropometric information to compute the distance using gold standard cameras. Lambrecht et al. [14] proposed three different algorithms to detect four events (HS, TO, FF, and HO) in stance phase. Three different set of kinematic data, obtained from the gold standard systems, were given as input to the algorithms. Data were collected from normal subjects during treadmill walking at three different speeds. Vertical GRF (from force plate) and vertical velocity of heel and toe markers were used for the ground truth. Each of the algorithms maintain a state machine to keep track of the sequence of events. Threshold for each type of data was computed based on the mean value of the preceding five events with addition of an offset. Despite reporting a promising result, the clinical significance of these methods was not verified. In addition, higher variability for HO event was observed. Usage of many kinematic features increases the complexity the methods.

5.3.2 Machine learning—based methods

A few studies have tried to segment gait cycle using machine learning algorithms. Miller et al. [15] mapped the event detection problem to binary classification (HS and TO) using a single hidden layer neural network (NN). Data were collected from different pathological populations using the gold standard cameras. Ground truth was generated from the force plate. A feature vector of size 15 was constructed using heel and toe marker data in sagittal plane. A sliding window was used to feed the NN on incremental basis. Each variable within the window was converted to a standard form which resulted a feature vector of size 315. Size of the feature space was reduced using principal component analysis (PCA). The NN was trained until the average root-mean-square error for training dataset reaches to a predefined value. In the output layer, the gait event was labeled by identifying local maxima in the time series of sigmoid activation function. NN was exhibited a competing performance in detecting gait event. Kidziński et al. [16] proposed a long short-term memory (LSTM) for event detection (HS and TO) in both normal and pathological populations. Data were collected from the gold standard system and ground truth was generated using manual annotation. Lower limb body positions and joint kinematics were used as the input vector corresponding to each frame. Misclassification error was computed using weighted binary cross-entropy. The network consists of a multilayer LSTM where fixed number of frames were used to provide the input. A sequence of different size windows of data frames was generated which were given as input to the system. The network mapped the input to n-dimensional time series. Subset of features, number of hidden layers, and number of horizontal layers were tuned during validation testing. Local maxima on the output time series were marked as the gait events. The proposed network performed better than the existing heuristic approaches. Zell et al. [20] segmented gait cycle into four phases as a part of a more generalized study where inverse dynamics was optimized using machine learning. Data were collected from normal subjects using the gold standard cameras. Force plate was used for the ground truth. Hermite polynomials were used to compute joint motion coefficients. Gait phases were estimated using multistage subclass control regression approach. Feature vector was constructed from motion coefficients and parameters that characterize the contact point of the subject with the ground. Support vector machine (SVM) (radial basis (RBF) kernel) was used to classify the phases. It is noteworthy that almost all of the above-mentioned studies have proposed their methods under an expensive system setup. Although Auvinet et al. [2] proposed a low-cost solution, the clinical significance of the method is questionable. A very few studies have reported about low-cost systems for vision-based gait event detection.

5.4 Kinect in gait cycle segmentation

In literature, many studies have obtained promising results in event detection or phase identification using high end cameras. But, a low-cost vision-based solution for this application is highly needed for many clinics, especially in developing countries. Kinect has emerged as one of the cheap and popular sensors in clinical domain. However, a very few studies have reported about gait event detection systems using Kinect sensor.

Auvinet et al. [2] computed HS events during treadmill walking of normal healthy subjects. Data were acquired from a Kinect sensor. Ground truth was generated from the gold standard cameras. The authors hypothesized that HS will occur when the distance between knees (longitudinal axis) would be maximum. Knee distance was computed in three different methods. In the first method, the gold standard system was used to estimate the distance using knee joints positions. In the second method, knee joints were localized on the depth map of Kinect using anthropometric data. Third method used the same anthropometric information to compute the distance using gold standard cameras. Chakraborty et al. [21] have proposed a multi-Kinect system to detect HS and TO events in treadmill gait. K-means algorithm was used to segment gait velocity time series. Then a heuristic approach was used to label the frames for stance and swing phases. The transition frames were marked as the gait events. Manual annotation was used to construct the ground truth. The authors have obtained a competing result. Despite increasing popularity of Kinect in clinical gait research, usage of this sensor in gait event detection is limited till now. More studies in this direction, especially using pathological population, are warranted.

5.5 Inertial sensor-based gait segmentation

5.5.1 Threshold-based methods

One of the initial approaches to address gait event detection is thresholding. It is a rule based method where the event transitions are identified by the model using the specified threshold values for the sensor readings [22—26]. In many studies, the rule-based methods are used to generate the ground truth data which is later used for training machine learning models [27]. However, the rule-based methods are inefficient in real-time scenario due to its inability to change the thresholds dynamically [28]. This instigated the use of machine learning over rule-based methods.

5.5.2 Machine intelligence-based methods

Application of machine learning techniques for gait event detection has seen a surge in the recent years. Bayesian inference model has been widely used for event detection [27,29,30]. Bayesian models use maximum aposterior probability rule to classify the events. In Ref. [27], the characteristics of the peaks and valleys during different gait events are modeled using a Bayesian network which was trained using ground truth data using thresholding. Martinez-Hernandez et al. in Refs. [29,30] used Bayesian

Inference system to calculate posterior probability with a decision and improve the prediction accuracy with an adaptive neuro-fuzzy system. Other than Bayesian, Hidden Markov Models (HMMs) have also been used for gait event detection because of its ability to detect transition between states, i.e., gait phases. The study [31] by Mannini et al. is one of the early studies which used left-right HMM for gait event detection. In Ref. [32], a distributed hierarchical HMM classifier is implemented and various HMM implementations achieved positive results. Fuzzy networks which are similar to threshold-based methods have also been used for this domain. In Ref. [33], a gait event system using HMM combined with Fuzzy NN for classification of events is implemented. In Ref. [29], an adaptive neuro-fuzzy system is used to predict the weighting parameter based on which final prediction is done. Artificial NNs with the advantage of nonparametric learning have been successful for the application of gait event detection as well [34,35]. In Ref. [34], the authors compared two rule-induced methods rough sets (RSs) and adaptive logic network (ALN) with rule-based thresholding method for gait phase detection from accelerometer signal. Although for detecting stance and swing, RS performed better than ALN; for five gait phase detection, ALN performed better then RS. Both [35,36] have used auto-regressive exogenous (ARX) models for learning the characteristics of gait phases. Apart from inertial sensors, other studies for electromyography (EMG) sensors [37,38] and vision sensors [15,39] have also used various NNs for segmenting gait phases from a signal. As in the domain of various signal processing applications, SVM have also been useful for classifying gait event from a gait sequence [40,41]. The application of SVM with RBF kernel is implemented in Ref. [41]. In Ref. [40], PCA is used as feature extraction technique, and the extracted features are used for classification using SVM. However, the application of SVM models has been limited in the recent studies for gait event detection using inertial sensors.

Recently with the advancement in the deep learning techniques, LSTM networks have been proven to be powerful tools for sequence prediction and classification. LSTM has thus been highly effective as gait event detection method [42,43]. A comparative analysis of LSTM and SVM is done in Ref. [42] whereas an LSTM−DNN network is implemented in Refs. [43,44]. Some of the relevant gait event detection methods are grouped together based on the methods discussed so far: Bayesian, SVM, HMM, Fuzzy, and LSTM in Table 5.2. The information regarding the type of sensors used and its placement in each study are also mentioned.

Table 5.2 Relevant studies of gait event detection methods using inertial sensors are classified into four categories: (1) Bayesian, (2) SVM, (3) HMM, (4) Fuzzy, and (5) LSTM.

	Experimental study	Sensor used	Sensor placement	Event detection method	Remarks
1	Lopez-Nava et al. [27]	Accelerometer	Ankle	Bayesian	A Bayesian classifier is used to detect heel strike and toe-off events
	Martinez-Hernandez et al. [29]	IMU	Thigh, shank, and foot	Adaptive neuro-fuzzy and Bayesian methods	A Bayesian perception system is developed to predict gait events. The final prediction is controlled by a weighting parameter using an adaptive neuro-fuzzy system
	Martinez-Hernandez et al. [30]	IMU	Thigh, shank, and foot	Adaptive Bayesian inference system	Gait event detection using Bayesian inference system is made adaptive with action-perception method
2	Taborri et al. [33]	IMU	Foot, shank, and thigh	Distributed HMM	Weighted decision from outputs of scalar Hidden Markov Model (HMM) is considered for classification
	Evans et al. [32]	IMU	Base of spine, thigh, shank, and foot	HMM + FNN hybrid model	A multilayer FNN is embedded within HMM for prediction of gait events sequence
	Chattopadhyay et al. [45]	IMU	Right shank	HMM	Symbolic aggregate approximation is used for observation sequence generation to train HMM model
3	Williamson et al. [34]	Accelerometer	Tibial crest	Adaptive logic network	Comparison between rough sets, adaptive logic network (ALN), and thresholding is performed for detecting gait phases
	Jung et al. [36]	IMU and foot pressure	Lower limb	Nonlinear ARX	A comparative analysis between an MLP network and nonlinear auto-regressive exogenous (ARX) model for detecting gait phases using IMU sensors in ROBIN-h1 exoskeleton robot
	Hesami et al. [35]	Inertial movement suit	Shank	Fuzzy ARX	A fuzzy ARX model is used for detecting gait phases

Table 5.2 Relevant studies of gait event detection methods using inertial sensors are classified into four categories: (1) Bayesian, (2) SVM, (3) HMM, (4) Fuzzy, and (5) LSTM.—*continued*

	Experimental study	Sensor used	Sensor placement	Event detection method	Remarks
4	Huang et al. [40]	Intelligent shoes, pedar insole system	Foot	SVM	Principal component analysis is used for feature selection and reduction. A support vector machine (SVM) with radial basis (RBF) kernel is used for gait event classification
	Xu et al. [41]	IMU	Shank	SVM	An SVM with RBF kernel is used to detect gait events to control a bionic knee exoskeleton
5	Ding et al. [42]	IMU	Shank	LSTM	A comparative analysis between long short-term memory (LSTM) and SVM is done using angle and angular velocity information
	Zhen et al. [44]	Tri axle accelerometer	Thigh, calf, and foot	LSTM	An LSTM-DNN network is used for gait data acquired for 0.78 m/s, 1.0 m/s, and 1.25 m/s
	Tan et al. [43]	Accelerometer	Shank	LSTM	An LSTM network trained and tested with walking and running gait dataset in treadmill and outdoor environment

5.6 Electromyography sensor-based gait segmentation

Machine learning techniques play an important role in gait segmentation into stance and swing phases for sEMG signals. A sequence of time-domain features is used to identify stance and swing phase using HMMs [46]. Gait event detection using sEMG sensor will support in development of assistive devices.

Nazmi et al. proposed gait phase classification using artificial NN-based on time domain features [37]. Wang et al. proposed EMG sensor-based method for understanding and analyzing variability in gait features through vector quantization networks [47]. Morbidoni et al. proposed sEMG sensor-based gait phase classification (stance/swing phase) using deep learning methods and also predicted foot-floor-contact signal in regular walking [48]. It provides a suitable performance in prediction and classification during ground level walking. Gait event detection from sEMG signals is an important step in clinical gait analysis. Machine learning techniques are quite successful in this field. Di Nardo et al. proposed an intra-subject approach using NN (multilayer perceptron architecture) for predicting gait events and classifying gait phases [49]. They also proposed a novel approach for identifying gait phases and detecting gait events using deep learning algorithms on sagittal knee angles using a single electrogoniometer [50].

Gait segmentation is a common procedure to divide the signal into segments. In preprocessing step, each segment is considered as one complete gait cycle, and the length of the segment matches with gait cycle length. Gait signal is cyclic in nature. A single gait cycle consists of all the eight phases. All these phases are repeated in consecutive gait cycles. For extracting a single gait cycle, we apply statistical approach based on signal autocorrelation.

5.6.1 Statistical methods

5.6.1.1 Autocorrelation

It is a process in which the signal is correlated with the lagged or delayed version of itself. The peaks of the autocorrelation are used to obtain the gait cycles from the processed EMG signals. Fig. 5.1 shows an example of autocorrelation output. The following equation is used for calculating autocorrelation (r_k).

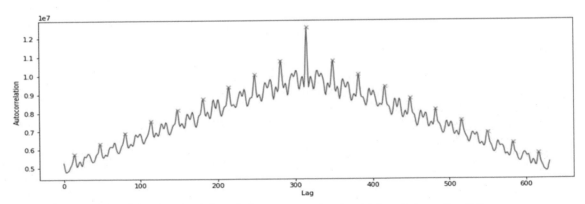

Figure 5.1 Autocorrelation of electromyography signal taken during gait activity.

$$r_k = \frac{\sum_{i=1}^{N-K}\left(X_i - \underline{X}\right)\left(X_{i+K} - \underline{X}\right)}{\sum_{i=1}^{N}\left(X_i - \underline{X}\right)^2}$$

k = Lag; 1, 2, ...
X_i = value of X at i^{th} instance.
$\underline{X}$= Mean of X.
N = Number of instances.

Autocorrelation is used to compare a signal with a time delayed version of itself. If a signal is periodic, then the signal will be perfectly correlated with a version of itself if the time-delay is an integer number of periods.

5.7 Summary

Gait cycle extraction and identification of gait events in a cycle is required for analyzing patterns of the signal during different phases. As the pattern of gait signals change with the sensor used for collecting the data, we described the methods applied for different motion sensors in this chapter. Most of these methods include the machine learning techniques used for identifying phase transition in signals and videos. Recent development in deep learning techniques has helped for automated pattern recognition to detect gait phases.

References

[1] B. Mariani, H. Rouhani, X. Crevoisier, K. Aminian, Quantitative estimation of foot-flat and stance phase of gait using foot-worn inertial sensors, Gait Posture 37 (2) (2013) 229–234, https://doi.org/10.1016/j.gaitpost.2012.07.012.
[2] E. Auvinet, F. Multon, C.E. Aubin, J. Meunier, M. Raison, Detection of gait cycles in treadmill walking using a kinect, Gait Posture (2015), https://doi.org/10.1016/j.gaitpost.2014.08.006.
[3] S. Ghoussayni, C. Stevens, S. Durham, D. Ewins, Assessment and validation of a simple automated method for the detection of gait events and intervals, Gait Posture 20 (3) (2004) 266–272, https://doi.org/10.1016/j.gaitpost.2003.10.001.
[4] F.A. Storm, C.J. Buckley, C. Mazzà, Gait event detection in laboratory and real life settings: accuracy of ankle and waist sensor based methods, Gait Posture 50 (2016) 42–46, https://doi.org/10.1016/j.gaitpost.2016.08.012.
[5] B.T. Smith, D.J. Coiro, R. Finson, R.R. Betz, J. McCarthy, Evaluation of force-sensing resistors for gait event detection to trigger electrical stimulation to improve walking in the child with cerebral palsy, IEEE Trans. Neural Syst. Rehabil. Eng. 10 (1) (2002) 22–29, https://doi.org/10.1109/TNSRE.2002.1021583.
[6] P. na Wei, R. Xie, R. Tang, C. Li, J. Kim, M. Wu, sEMG based gait phase recognition for children with spastic cerebral palsy, Ann. Biomed. Eng. 47 (1) (2019) 223–230, https://doi.org/10.1007/s10439-018-02126-8.

[7] J. Taborri, E. Scalona, E. Palermo, S. Rossi, P. Cappa, Validation of inter-subject training for hidden Markov models applied to gait phase detection in children with cerebral palsy, Sensors 15 (9) (2015) 24514–24529.

[8] R.T. Lauer, B.T. Smith, D. Coiro, R.R. Betz, J. McCarthy, Feasibility of gait event detection using intramuscular electromyography in the child with cerebral palsy, Neuromodulation 7 (3) (2004) 205–213, https://doi.org/10.1111/j.1094-7159.2004.04200.x.

[9] J. Taborri, E. Palermo, S. Rossi, P. Cappa, Gait partitioning methods: a systematic review, Sensors 16 (1) (2016) 40–42, https://doi.org/10.3390/s16010066.

[10] F. Attal, Y. Amirat, A. Chibani, S. Mohammed, Automatic recognition of gait phases using a multiple-regression hidden Markov model, IEEE/ASME Trans. Mechatronics 23 (4) (2018) 1597–1607.

[11] J.A. Zeni Jr., J.G. Richards, J.S. Higginson, Two simple methods for determining gait events during treadmill and overground walking using kinematic data, Gait Posture 27 (4) (2008) 710–714.

[12] C.M. O'Connor, S.K. Thorpe, M.J. O'Malley, C.L. Vaughan, Automatic detection of gait events using kinematic data, Gait Posture 25 (3) (2007) 469–474.

[13] Y. Tang, Z. Li, H. Tian, J. Ding, B. Lin, Detecting toe-off events utilizing a vision-based method, Entropy 21 (4) (2019) 329.

[14] S. Lambrecht, A. Harutyunyan, K. Tanghe, M. Afschrift, J. De Schutter, I. Jonkers, Real-time gait event detection based on kinematic data coupled to a biomechanical model, Sensors 17 (4) (2017), https://doi.org/10.3390/s17040671.

[15] A. Miller, Gait event detection using a multilayer neural network, Gait Posture 29 (4) (2009) 542–545.

[16] Ł. Kidziński, S. Delp, M. Schwartz, Automatic real-time gait event detection in children using deep neural networks, PLoS One 14 (1) (2019) e0211466.

[17] K. Tanghe, F. De Groote, D. Lefeber, J. De Schutter, E. Aertbelien, Gait trajectory and event prediction from state estimation for exoskeletons during gait, IEEE Trans. Neural Syst. Rehabil. Eng. (2020), https://doi.org/10.1109/TNSRE.2019.2950309.

[18] M. Lempereur, et al., A new deep learning-based method for the detection of gait events in children with gait disorders: proof-of-concept and concurrent validity, J. Biomech. (2020), https://doi.org/10.1016/j.jbiomech.2019.109490.

[19] S. Yu, D. Tan, T. Tan, A Framework for Evaluating the Effect of View Angle, Clothing and Carrying Condition on Gait Recognition, 2006, https://doi.org/10.1109/ICPR.2006.67.

[20] P. Zell, B. Rosenhahn, Learning inverse dynamics for human locomotion analysis, Neural Comput. Appl. (2020), https://doi.org/10.1007/s00521-019-04658-z.

[21] S. Chakraborty, A. Nandy, An unsupervised approach for gait phase detection, 4th Int. Conf. Comput. Intell. Networks, CINE 2020 (2020) 1–5, https://doi.org/10.1109/CINE48825.2020.234396.

[22] A.M. Sabatini, C. Martelloni, S. Scapellato, F. Cavallo, Assessment of walking features from foot inertial sensing, IEEE Trans. Biomed. Eng. 52 (3) (2005) 486–494.

[23] I.P.I. Pappas, M.R. Popovic, T. Keller, V. Dietz, M. Morari, A reliable gait phase detection system, IEEE Trans. Neural Syst. Rehabil. Eng. 9 (2) (2001) 113–125.

[24] H.M. Franken, W. de Vries, P.E. Veltink, G. Baardman, H.B.K. Boom, State detection during paraplegic gait as part of a finite state based controller, in:

Proceedings of the 15th Annual International Conference of the IEEE
Engineering in Medicine and Biology Society, 1993, pp. 1322—1323.

[25] E.D. Ledoux, Inertial sensing for gait event detection and transfemoral
prosthesis control strategy, IEEE Trans. Biomed. Eng. 65 (12) (2018)
2704—2712.

[26] N. Pinkam, I. Nilkhamhang, Wireless smart shoe for gait analysis with
automated thresholding using PSO, in: 2013 10th International Conference
on Electrical Engineering/Electronics, Computer, Telecommunications and
Information Technology, 2013, pp. 1—6.

[27] I.H. López-Nava, A. Muñoz-Meléndez, A.I. Pérez Sanpablo, A. Alessi
Montero, I. Quiñones Urióstegui, L. Núñez Carrera, Estimation of temporal
gait parameters using Bayesian models on acceleration signals, Comput.
Methods Biomech. Biomed. Eng. 19 (4) (2016) 396—403.

[28] M.M. Skelly, H.J. Chizeck, Real-time gait event detection for paraplegic FES
walking, IEEE Trans. Neural Syst. Rehabil. Eng. 9 (1) (2001) 59—68.

[29] U. Martinez-Hernandez, A. Rubio-Solis, G. Panoutsos, A.A. Dehghani-Sanij,
A combined adaptive neuro-fuzzy and bayesian strategy for recognition and
prediction of gait events using wearable sensors, in: 2017 IEEE International
Conference on Fuzzy Systems (FUZZ-IEEE), 2017, pp. 1—6.

[30] U. Martinez-Hernandez, A.A. Dehghani-Sanij, Adaptive Bayesian inference
system for recognition of walking activities and prediction of gait events
using wearable sensors, Neural Networks 102 (2018) 107—119.

[31] A. Mannini, A.M. Sabatini, A hidden Markov model-based technique for gait
segmentation using a foot-mounted gyroscope, in: 2011 Annual
International Conference of the IEEE Engineering in Medicine and Biology
Society, 2011, pp. 4369—4373.

[32] R.L. Evans, D.K. Arvind, Detection of gait phases using orient specks for
mobile clinical gait analysis, in: 2014 11th International Conference on
Wearable and Implantable Body Sensor Networks, 2014, pp. 149—154.

[33] J. Taborri, S. Rossi, E. Palermo, F. Patanè, P. Cappa, A novel HMM
distributed classifier for the detection of gait phases by means of a wearable
inertial sensor network, Sensors 14 (9) (2014) 16212—16234.

[34] R. Williamson, B.J. Andrews, Gait event detection for FES using
accelerometers and supervised machine learning, IEEE Trans. Rehabil. Eng.
8 (3) (2000) 312—319.

[35] S. Hesami, F. Naghdy, D.A. Stirling, H.C. Hill, Application of Fuzzy NARX to
Human Gait Modelling and Identification, 2008.

[36] J.-Y. Jung, W. Heo, H. Yang, H. Park, A neural network-based gait phase
classification method using sensors equipped on lower limb exoskeleton
robots, Sensors 15 (11) (2015) 27738—27759.

[37] N. Nazmi, M.A.A. Rahman, S.-I. Yamamoto, S.A. Ahmad, Walking gait event
detection based on electromyography signals using artificial neural
network, Biomed. Signal Process. Control 47 (2019) 334—343.

[38] R.T. Lauer, B.T. Smith, R.R. Betz, Application of a neuro-fuzzy network for
gait event detection using electromyography in the child with cerebral
palsy, IEEE Trans. Biomed. Eng. 52 (9) (2005) 1532—1540.

[39] M. Ye, C. Yang, V. Stankovic, L. Stankovic, S. Cheng, Gait phase
classification for in-home gait assessment, in: 2017 IEEE International
Conference on Multimedia and Expo (ICME), 2017, pp. 1524—1529.

[40] B. Huang, M. Chen, X. Shi, Y. Xu, Gait event detection with intelligent
shoes, in: 2007 International Conference on Information Acquisition, 2007,
pp. 579—584.

[41] D. Xu, X. Liu, Q. Wang, Knee exoskeleton assistive torque control based on real-time gait event detection, IEEE Trans. Med. Robot. Bionics 1 (3) (2019) 158–168.

[42] Z. Ding, et al., The real time gait phase detection based on long short-term memory, in: 2018 IEEE Third International Conference on Data Science in Cyberspace (DSC), 2018, pp. 33–38.

[43] H.X. Tan, N.N. Aung, J. Tian, M.C.H. Chua, Y.O. Yang, Time series classification using a modified LSTM approach from accelerometer-based data: a comparative study for gait cycle detection, Gait Posture 74 (2019) 128–134.

[44] T. Zhen, L. Yan, P. Yuan, Walking gait phase detection based on acceleration signals using LSTM-DNN algorithm, Algorithms 12 (12) (2019) 253.

[45] S. Chattopadhyay, A. Nandy, Human gait modelling using hidden Markov model for abnormality detection, in: TENCON 2018-2018 IEEE Region 10 Conference, 2018, pp. 623–628.

[46] M. Meng, Q. She, Y. Gao, Z. Luo, EMG signals based gait phases recognition using hidden Markov models, in: The 2010 IEEE International Conference on Information and Automation, 2010, pp. 852–856.

[47] J. Wang, T. Zielińska, Gait features analysis using artificial neural networks: testing the footwear effect, Acta Bioeng. Biomech. 19 (1) (2017) 17–32.

[48] C. Morbidoni, A. Cucchiarelli, S. Fioretti, F. Di Nardo, A deep learning approach to EMG-based classification of gait phases during level ground walking, Electronics 8 (8) (2019) 894.

[49] F. Di Nardo, C. Morbidoni, G. Mascia, F. Verdini, S. Fioretti, Intra-subject approach for gait-event prediction by neural network interpretation of EMG signals, Biomed. Eng. Online 19 (1) (2020) 1–20.

[50] F. Di Nardo, C. Morbidoni, A. Cucchiarelli, S. Fioretti, Recognition of gait phases with a single knee electrogoniometer: a deep learning approach, Electronics 9 (2) (2020) 355.

6

Methodologies for vision-based automatic pathological gait detection

6.1 Introduction

Quantitative gait assessment has dramatically uplifted the clinical significance of gait. Diagnostic systems or sometimes referred as gait abnormality detection systems heavily relied on quantification of salient gait parameters or features. Such diagnostic systems are gradually becoming very popular in clinics. It helps the doctor in assessing the improvement of a patient after a therapeutic intervention. Minor change in gait which is not visually observable could be significant to assess the impact of an intervention. Doctor can recommend to continue the exercise or can prescribe a different exercise depending on the assessment of the progress [1]. In the realm of quantitative gait analysis computational intelligence (CI)-based diagnosis, especially machine learning—based techniques, have now become very popular for its ability to handle high dimensional nonlinear data [2]. These systems provide more precise, robust, and fast solutions. Detection of abnormality is performed by classifying the gait pattern between normal and abnormal populations. No manual intervention is required to analyze the features. Algorithms automatically learn the nonlinear relationship by minimizing the error between the predicted and actual outcome. Threshold between the classes is automatically adjusted by those algorithms based on the available dataset. Automatic learning ability from data makes it promising for gait analysis. Numerous works have been performed to devise automatic gait diagnostic system using CI. Sensor plays a vital role in providing data to the intelligent algorithms. Wearable sensor-based systems suffer from its generic problem. It makes the participants uncomfortable during walking which may alter the gait pattern [3]. On the other hand,

Modern Methods for Affordable Clinical Gait Analysis. https://doi.org/10.1016/B978-0-323-85245-6.00006-0

nonwearable devices, especially vision-based sensors, carry advantage on this aspect. But most of the studies in state-of-the-art used expensive vision-based systems (e.g., Vicon, Qualisys, etc.) to devise an automatic gait abnormality detection system. A very few studies have reported affordable and automatic vision-based gait diagnostic systems mainly using Kinect sensor skeletal data stream. This chapter first describes the prevailing gait abnormality detection techniques and then it demonstrates some affordable gait diagnostic systems using Kinect sensor.

6.2 Gait detection techniques

Normally, gait assessment is performed in two different ways: qualitative and quantitative. In qualitative technique, pathologist or clinical staff evaluate the gait pattern on a predefined scale. This technique is simple, fast, and does not require any complex data analysis [4]. It can further be divided into the following two categories.

6.2.1 Observation-based

Here the pathologist asks the patient to perform some specific type of movement activities and after visual inspection assign a score depending on which the subjects' gaits are classified. The observation may be direct or indirect (e.g., simple video recording). Clinicians often use some task-specific rating scales or tests, such as timed get up and go test [5], gait abnormality rating scale [6], physical rating scale [7], observational gait score [8], etc., based on which overall score is computed.

6.2.2 Question answer-based

In this case, some questions are setup by the clinicians and the patients are asked to fill it up [9,10]. Sometimes the clinical staff himself/herself fill up the questions after taking telephonic interview of the patients [11].

In spite of simplicity, this technique suffers from high subjectivity [12]. The overall assigned score and consequently the diagnosis is highly dependent on the knowledge and experience level of the concerned clinical staff. Diagnostic result may vary with respect to different pathologists. Hence, the reliability of diagnosis is very low in this technique. On the other hand, quantitative methods depend on extraction of intrinsic features or auto-recognition of gait pattern from the time series data acquired from different sensors. Features are analyzed statistically or computationally and compared with the normal population to provide a clinical decision. Subtle deviation in gait pattern, which is not possible to observe in bare eye, can be tracked in quantitative analysis. This technique is highly objective, and more precise and reliable than the previous one [4]. However, there are some ongoing criticisms and debates about the clinical relevance and practicality of quantitative gait analysis, such as availability of standard normal population data [13], effect of laboratory environment on gait pattern [13,14], trained and experienced personnel to make the sensor setup [4], proper interpretation of large scale clinical data in some cases [4], etc. Despite aforementioned issues, quantitative analysis has now become an accepted clinical means to document and evaluate a persons' gait [14]. In the realm of quantitative gait analysis, two type of techniques are mainly used for data analysis: statistics-based and CI-based.

6.2.3 Statistics-based

In this technique, some gait parameters are extracted from the gait signal of both normal and pathological populations. Then the difference between the gait patterns is computed by performing a set of statistical tests (e.g., paired t-tests, Kolmogorov–Smirnov test, Mann–Whitney U test, etc.) on the gait parameters [15–21]. The deviation of gait parameters is used to identify the abnormality. Here human intervention is required to construct the threshold based on which decision on classification is made.

6.2.4 Computational intelligence-based

This technique includes supervised, unsupervised, fuzzy, evolutionary algorithms to hybrid learning. However, the first two paradigms, which fall under the category of machine learning, have been deployed most frequently. The outline of this procedure is demonstrated next. Preprocessed gait signal or extracted features are given as input to the learning algorithms which map the input features to the output class labels. Fig. 6.1 describes the process.

After acquisition of data from sensors, some preprocessing algorithms are applied to remove noise and artifacts from the signal. Then some salient features are extracted to construct the feature vector. Sometimes feature selection algorithms are used to reduce high dimensionality of features. Instead of extracting features manually, some deep learning algorithms may be used which extract features (deep features) automatically. Data

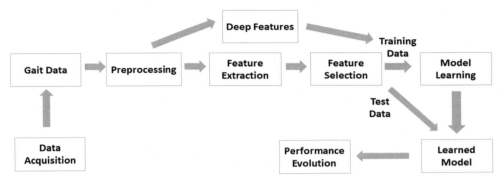

Figure 6.1 Generic flowchart of computational intelligence-based methods.

are divided into training and test set. Sometimes training set is further divided into two parts; training and validation sets. Validation data are used to determine the model's goodness of fit. The model is trained using the training set and evaluated on the test set. During training, the threshold between the classes is automatically adjusted by the learning algorithm based on optimization principle. Hence, the gait abnormality detection systems based on the above technique is sometimes referred as automatic gait diagnostic systems.

The overall cost of this automatic gait diagnostic system highly depends on the data acquisition sensors. The prevailing most popular vision-based sensors, sometimes called the *gold standard* sensors, are expensive and unaffordable for most of the clinics, especially in developing countries. Usually, a system comprising of four to eight such cameras are recommended for a gait lab [22], which cost around 30,000 USD [23]. Again, its complex calibration technique and software handling procedure requires expertise. Hence, as an alternative, an effort toward affordable gait analysis has been observed for a few years. Recently, Microsoft Kinect, a depth sensing camera, has become very popular in clinical gait analysis due to its portability, unobtrusive data acquisition property, and especially inexpensiveness (≈ 100 USD) [24].

A very few studies have reported an affordable and automatic vision-based gait diagnostic system using Kinect sensor skeletal data stream. Mostly skeletal data stream was used for diagnosis for its simplicity to obtain joint positional information. Those studies have obtained a promising and competing results. Some discussions on this aspect are provided in the subsequent section.

6.3 Automatic diagnostic systems using Kinect

A few studies have tried to build cost-effective vision-based automatic gait diagnostic systems. Generally, flat surface overground experiment is performed for gait abnormality detection using real patients. Sensor is placed in front of the subject such that all body joints are clearly visible and tracked easily. Subjects are asked to walk at self-selected speed toward the sensor during which data are collected. Some popular gait diagnostic systems are described below:

Bei et al. [25] proposed a gait detection system using kinematic and SPT features. Data were collected from a Kinect v2 sensor (skeleton data stream) which were placed in front of the subjects. A three step procedure was recommended to remove noise from the skeletal data stream. In step 1, the frames where some important joint positions were missing or dislocated were marked as the corrupted frames. In next step, three-dimensional data of each joints were scaled to a common factor. In step 3, a novel smoothing algorithm was proposed. A gait symmetry metric based on lower limb joint angles was proposed. Step length and stride time were computed using zero-crossing detection method [26]. Using different combinations of those three features a comparative analysis between different classifiers (i.e., k-means, Bayesian classifier, SVM, and kNN) was performed to detect the abnormal gait. The impact of gender and age on classification was also assessed. The authors obtained 95% accuracy.

Dolatabadi et al. [27] proposed an automated gait diagnosis system using two Kinect sensors. The sensors were placed at two opposite end of walking track. Data were collected from only one sensor at a time when the subject was walking toward it. Joint angles from 12 body joints and velocity as well as acceleration of spine base were taken as features. Two classifiers, i.e., kNN and Gaussian process latent variable model [28] were trained using nested cross validation. Inner loop was used for hyperparameter tuning whereas in outer loop 10-fold cross validation was performed. In addition, a feature level analysis was performed to investigate the importance of subset of features in gait classification. The two classifiers were reported to provide a satisfactory result for different metrics.

Li et al. [29] constructed a gait diagnostic system for Parkinson and hemiplegic patients. They proposed a speed invariant gait descriptor which was represented as a covariance matrix. The normalized skeleton data were segmented into overlapping time windows. For each window, joints relative positions and speed information were encoded in a covariance matrix. The classifier (K-nearest neighbors) learned the distance between the covariance matrices to model gait. The authors reported 91.1% accuracy for abnormality detection.

As mentioned in the above studies and some in few other studies like Ref. [30–32], a single Kinect architecture was proposed for data collection. But, due to limited FoV of Kinect, the walking track length becomes too small to become clinically relevant, i.e., 10 m or 8 m of length. As an alternative, recently, few studies have tried construct a clinically relevant walking track in a multi-Kinect setup based on which gait diagnostic systems were proposed.

6.4 Gait diagnosis in multi-Kinect architecture

A multi-Kinect setup could provide a long walking track. Chakraborty et al. [33] have proposed an architecture using three Kinects v2 (see Fig. 6.2) which provide an effective walking length of 8 m (approx.). A client-server protocol was used to control the Kinects. An overlapped tracking volume of 1 m (approx.) between two successive sensors was constructed which was empirically estimated to be sufficient for the next sensor to recognize a person's body and start skeleton tracking. The client-server architecture followed the same protocol like Müller et al. [34] to record, combine, and process body point data from multiple Kinects. Authors diagnosed Cerebral Palsy gait using the above structure. The

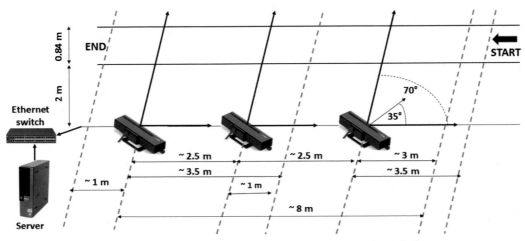

Figure 6.2 A multi-Kinect architecture for over-ground gait [33].

combined Kinect skeleton data were observed to have different outliers due to incorporation of noise. A data-driven algorithm was proposed to remove those outliers. In addition, a gait speed invariant handcrafted feature was introduced and tested for that population. Different classifiers were used to test the utility of the proposed algorithm and the feature. The study reported 98% (approx.) of accuracy using a single feature and support vector machines.

The above architecture with some modification was used in Chakraborty et al. [35] to construct a state-space model where abnormality of gait pattern was modeled by quantifying feature uncertainty. Temporal evolution of gait pattern was conceptualized using Cardiff classifier [36]. The authors also proposed two abnormality indices.

Another work by Chakraborty et al. [37] detected gait abnormality during treadmill gait experiment. Normal subjects were asked to simulate pathological gait. A novel sensor setup was proposed using three Kinects whose effective FoV was more than 180 degrees (see Fig. 6.3). Sensors data were fused using FusionKit software [38] which uses client-server architecture to calibrate data. Joint position data were collected from the fused skeleton. Lower limb velocity was used as input feature for the classifier which detected abnormal gait with 93% of accuracy.

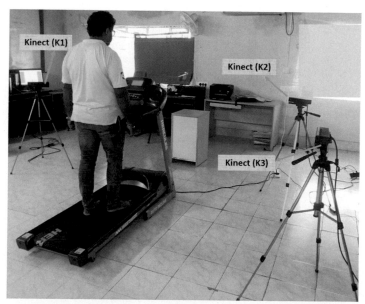

Figure 6.3 A multi-Kinect architecture for treadmill gait [37].

In spite of some promising results as mentioned above, research on gait abnormality detection in multi-Kinect environment is still in its infant state. More investigation in this domain is warranted using different pathological populations which will be cost-effective as well as clinically significant.

6.5 Summary

This chapter demonstrates the necessity of gait abnormality detection systems and different prevailing techniques for it. A comparative analysis between the qualitative and quantitative gait assessment methods was also presented. Finally, some low-cost solutions for gait abnormality detection using CI-based techniques were explained. This chapter might be a helpful guide for the clinicians to establish an affordable automatic gait assessment system.

References

[1] J. Figueiredo, C.P. Santos, J.C. Moreno, Automatic recognition of gait patterns in human motor disorders using machine learning: a review, Med. Eng. Phys. 53 (2018) 1–12.

[2] D.T.H. Lai, R.K. Begg, M. Palaniswami, Computational intelligence in gait research: a perspective on current applications and future challenges, IEEE Trans. Inf. Technol. Biomed. 13 (5) (2009) 687–702, https://doi.org/10.1109/TITB.2009.2022913.

[3] S. Hagler, D. Austin, T.L. Hayes, J. Kaye, M. Pavel, Unobtrusive and ubiquitous in-home monitoring: a methodology for continuous assessment of gait velocity in elders, IEEE Trans. Biomed. Eng. 57 (4) (2010) 813.

[4] F. Moissenet, S. Armand, Qualitative and quantitative methods of assessing gait disorders, in: Orthopedic Management of Children with Cerebral Palsy: A Comprehensive Approach, 2015.

[5] S. Richardson, The timed 'up & go': a test of basic functional mobility for frail elderly persons, J. Am. Geriatr. Soc. (1991), https://doi.org/10.1111/j.1532-5415.1991.tb01616.x.

[6] L. Wolfson, R. Whipple, P. Amerman, J.N. Tobin, Gait assessment in the elderly: a gait abnormality rating scale and its relation to falls, J. Gerontol. (1990), https://doi.org/10.1093/geronj/45.1.M12.

[7] K.G.B. Maathuis, C.P. Van Der Schans, A. Van Iperen, H.S. Rietman, J.H.B. Geertzen, Gait in children with cerebral palsy: observer reliability of physician rating scale and Edinburgh visual gait analysis interval testing scale, J. Pediatr. Orthop. (2005), https://doi.org/10.1097/01.bpo.0000151061.92850.74.

[8] A.H. Mackey, G.L. Lobb, S.E. Walt, N.S. Stott, Reliability and validity of the Observational Gait Scale in children with spastic diplegia, Dev. Med. Child Neurol. (2003), https://doi.org/10.1111/j.1469-8749.2003.tb00852.x.

[9] K. Delbaere, K. Hauer, S.R. Lord, Evaluation of the incidental and planned questionnaire (IPEQ) for older people, Br. J. Sports Med. (2010), https://doi.org/10.1136/bjsm.2009.060350.

[10] K. Hauer, et al., Validation of the falls efficacy scale and falls efficacy scale international in geriatric patients with and without cognitive impairment: results of self-report and interview-based questionnaires, Gerontology (2010), https://doi.org/10.1159/000236027.

[11] H. Wagner, H. Melhus, R. Gedeborg, N.L. Pedersen, K. Michaëlsson, Simply ask them about their balance - future fracture risk in a nationwide cohort study of twins, Am. J. Epidemiol. (2009), https://doi.org/10.1093/aje/kwn379.

[12] D. Hamacher, N.B. Singh, J.H. Van Dieën, M.O. Heller, W.R. Taylor, Kinematic measures for assessing gait stability in elderly individuals: a systematic review, J. R. Soc. Interface 8 (65) (2011) 1682–1698, https://doi.org/10.1098/rsif.2011.0416.

[13] D. Rosenbaum, M. Brandes, Qualitative and Quantitative Aspects of Movement: The Discrepancy between Clinical Gait Analysis and Activities of Daily Life, 2008.

[14] S.A. Gard, Use of quantitative gait analysis for the evaluation of prosthetic walking performance, J. Prosthetics Orthot. (2006), https://doi.org/10.1097/00008526-200601001-00011.

[15] C.-F. Chang, et al., Balance control during level walking in children with spastic diplegic cerebral palsy, Biomed. Eng. Appl. Basis Commun. 23 (06) (2011) 509–517.

[16] C.J. Kim, Y.M. Kim, D.D. Kim, Comparison of children with joint angles in spastic diplegia with those of normal children, J. Phys. Ther. Sci. 26 (9) (2014) 1475–1479, https://doi.org/10.1589/jpts.26.1475.

[17] T.-M. Wang, H.-P. Huang, J.-D. Li, S.-W. Hong, W.-C. Lo, T.-W. Lu, Leg and joint stiffness in children with spastic diplegic cerebral palsy during level walking, PLoS One 10 (12) (2015) e0143967.

[18] S. Tsukagoshi, et al., Noninvasive and quantitative evaluation of movement disorder disability using an infrared depth sensor, J. Clin. Neurosci. 71 (xxxx) (2020) 135–140, https://doi.org/10.1016/j.jocn.2019.08.101.

[19] J. Latorre, C. Colomer, M. Alcañiz, R. Llorens, Gait analysis with the Kinect v2: normative study with healthy individuals and comprehensive study of its sensitivity, validity, and reliability in individuals with stroke, J. Neuroeng. Rehabil. 16 (1) (2019) 1–11, https://doi.org/10.1186/s12984-019-0568-y.

[20] A.P. Rocha, H. Choupina, J.M. Fernandes, M.J. Rosas, R. Vaz, J.P.S. Cunha, Kinect v2 based system for Parkinson's disease assessment, Proc. Ann. Int. Conf. IEEE Eng. Med. Biol. Soc. EMBS 2015-November (2015) 1279–1282, https://doi.org/10.1109/EMBC.2015.7318601.

[21] A. Prochazka, M. Schatz, O. Tupa, M. Yadollahi, O. Vysata, M. Walls, The MS kinect image and depth sensors use for gait features detection, IEEE Int. Conf. Image Process. ICIP 2014 (2014) 2271–2274, https://doi.org/10.1109/ICIP.2014.7025460.

[22] C. Prakash, R. Kumar, N. Mittal, Recent developments in human gait research: parameters, approaches, applications, machine learning techniques, datasets and challenges, Artif. Intell. Rev. 49 (1) (2018), https://doi.org/10.1007/s10462-016-9514-6.

[23] S. Chen, J. Lach, B. Lo, G.Z. Yang, Toward pervasive gait analysis with wearable sensors: a systematic review, IEEE J. Biomed. Heal. Inform. 20 (6) (2016) 1521–1537, https://doi.org/10.1109/JBHI.2016.2608720.

[24] T. Guzsvinecz, V. Szucs, C. Sik-Lanyi, Suitability of the kinect sensor and leap motion controller-A literature review, Sensors 19 (5) (2019), https://doi.org/10.3390/s19051072.

[25] S. Bei, Z. Zhen, Z. Xing, L. Taocheng, L. Qin, Movement disorder detection via adaptively fused gait analysis based on kinect sensors, IEEE Sensor. J. 18 (17) (2018) 7305–7314, https://doi.org/10.1109/JSEN.2018.2839732.

[26] R.J. Aliaga, Real-time estimation of zero crossings of sampled signals for timing using cubic spline interpolation, IEEE Trans. Nucl. Sci. (2017), https://doi.org/10.1109/TNS.2017.2721103.

[27] E. Dolatabadi, B. Taati, A. Mihailidis, An automated classification of pathological gait using unobtrusive sensing technology, IEEE Trans. Neural Syst. Rehabil. Eng. 25 (12) (2017) 2336–2346.

[28] J.M. Wang, D.J. Fleet, A. Hertzmann, Gaussian process dynamical models, in: Advances in Neural Information Processing Systems, 2005.

[29] Q. Li, et al., Classification of gait anomalies from kinect, Vis. Comput. 34 (2) (2018) 229–241, https://doi.org/10.1007/s00371-016-1330-0.

[30] C. Urcuqui, et al., Exploring machine learning to analyze Parkinson's disease patients, Proc. 2018 14th Int. Conf. Semant. Knowl. Grids SKG (2018) 160–166, https://doi.org/10.1109/SKG.2018.00029.

[31] T.N. Nguyen, H.H. Huynh, J. Meunier, Skeleton-based abnormal gait detection, Sensors 16 (11) (2016) 1–13, https://doi.org/10.3390/s16111792.

[32] M. Khokhlova, C. Migniot, A. Morozov, O. Sushkova, A. Dipanda, Normal and pathological gait classification LSTM model, Artif. Intell. Med. 94 (2019) 54–66.

[33] S. Chakraborty, A. Nandy, Automatic diagnosis of cerebral palsy gait using computational intelligence techniques: a low-cost multi-sensor approach, IEEE Trans. Neural Syst. Rehabil. Eng. (2020), https://doi.org/10.1109/TNSRE.2020.3028203.

[34] B. Müller, W. Ilg, M.A. Giese, N. Ludolph, Validation of enhanced kinect sensor based motion capturing for gait assessment, PLoS One 12 (4) (2017) e0175813.

[35] S. Chakraborty, N. Thomas, A. Nandy, Gait abnormality detection in people with cerebral palsy using an uncertainty-based state-space model, in: International Conference on Computational Science, 2020, pp. 536–549.

[36] P.R. Biggs, G.M. Whatling, C. Wilson, C.A. Holt, Correlations between patient-perceived outcome and objectively-measured biomechanical change following Total Knee Replacement, Gait Posture 70 (March 2018) (2019) 65–70, https://doi.org/10.1016/j.gaitpost.2019.02.028.

[37] S. Chakraborty, R. Mishra, A. Dwivedi, T. Das, A. Nandy, A low-cost pathological gait detection system in multi-kinect environment, in: Springer Proceedings in Physics, vol. 249, 2020, https://doi.org/10.1007/978-981-15-6467-3_13.

[38] M. Rietzler, F. Geiselhart, J. Thomas, E. Rukzio, FusionKit: a generic toolkit for skeleton, marker and rigid-body tracking, in: Proceedings of the 8th ACM SIGCHI Symposium on Engineering Interactive Computing Systems, 2016, pp. 73–84.

Pathological gait pattern analysis using inertial sensor

7.1 Introduction

Damage to any part of central nervous system may result in muscle stiffness, which limits normal movement. Gait analysis in the rehabilitation process allows people with neurological disorders to make significant decisions, including modifications to physical therapy routines. Three-dimensional motion capture (mocap) method is commonly used for measuring gait movement and calculation of various statistical parameters (for example, cadence and stride length) in a traditional clinical gait analysis system to diagnose neurologically disordered patients [29–31]. These analysis systems are considered as the most standard in the literature. However, there are some limitations of such systems, such as they have extremely high cost and fixed installation set-up. To overcome these limitations, InertiaLocoGraphy (ILG) and Inertial sensor based quantitative gait analysis can be used [32]. One of the advantages of using inertial sensors is that it can be performed even in an un-constrained environment and installation setup is easy. Recent development of MEMS technology has helped inertial sensors to be efficiently small and concise, making it comfortable for patients to wear during walking. Usage of inertial sensors for pathological gait assessment is still at preliminary stage, however has shown impressive progress. Inertial sensor has been used for assessing various neurological diseases such as Parkinson's disease [1–5], cerebral palsy [6–8], post-stroke subjects [9–11], Huntington's disease [9], etc. In the following sections, we discuss the steps for pathological gait assessment using inertial sensor. In addition to that, an example study with sample data is also demonstrated.

Modern Methods for Affordable Clinical Gait Analysis. https://doi.org/10.1016/B978-0-323-85245-6.00011-4

7.2 Data collection

7.2.1 Sensor specifications: Sparkfun 9DoF Razor inertial measurement unit

To develop a cost-effective gait assessment system, Sparkfun 9DoF Razor inertial measurement unit (IMU) sensor, an affordable IMU sensor, is used for the wearable prototype, shown in Fig. 7.1. This sensor contains MPU-9250 9DoF sensor, one of the latest sensors introduced by InvenSense, and a SAMD21 microprocessor which enables the IMU sensor to be reprogrammable. MPU-9250 is a combination of two chips: MPU-6500 which has a 3-axis gyroscope and a 3-axis accelerometer, and AK8963, a 3-axis magnetometer. MPU-9250 has the advantages such as low-power consumption and comparatively smaller size which makes it suitable for the wearable prototype. In addition to that, its gyroscope sensor has 3× better performance than previous models. The onboard SAMD21 microprocessor is an arduino-compatible microcontroller. The IMU sensor comes with a micro USB slot which can be used to store the digitalized values of the 3-axis gyroscope sensor, 3-axis accelerometer, and 3-axis magnetometer sensor. A single cell lithium-polymer battery can be used to power the sensor when in mobile. The battery can also be easily recharged with the help of USB supply. The sensor is programmed to capture the data with 100Hz frequency.

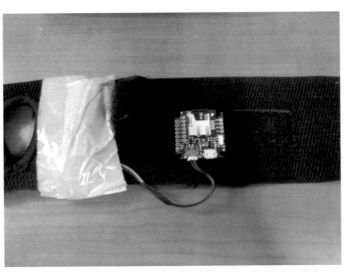

Figure 7.1 A prototype model of wearable sensor using Sparkfun 9DoF Razor inertial measurement unit sensor.

7.2.2 Sensor calibration

One of the major reasons why inertial sensors are not generally used for clinical practice is because these sensors are prone to error and can drift depending on noise, nonlinearity, and environmental factors. This requires IMU sensors to be calibrated to rectify the drifting of measured values. In the calibration process, the generated outputs are compared with a reference system and modeling the factors that will make the outputs agree with the reference system. The conventional methods for calibration use mechanical platforms and rate tables which are very expensive and are performed in controlled environment, thus not feasible for MEMU IMU sensor calibration. Some researchers have modeled the sensor error mathematically.

7.2.2.1 Error model for accelerometer

If there is a small misalignment between the nonorthogonal sensor axes and the orthogonal body axes, the force measured by the accelerometer in sensor (nonorthogonal) coordinates can be transformed into orthogonal body coordinate axes as

$$A^o = \left(T^o_s\right)^{-1} A^s$$

where $T^o{}_s$ is,

$$\left(T^o_s\right)^{-1} = \begin{bmatrix} 1 & -\theta_{yz} & \theta_{zy} \\ \theta_{xz} & 1 & -\theta_{zx} \\ -\theta_{xy} & \theta_{yx} & 1 \end{bmatrix}$$

where A^s is the force measured in the accelerometer's coordinates, A^o presents force in orthogonal body coordinates and rotation matrix which describes the correction in rotation is depicted as T^o_s.

Assuming that the sensor axis x_s aligns with the body axis xo and that the axis y_s lies in the plane formed by x^o and y^o axes, the angles θ_{xz}, θ_{xy}, and θ_{yx} becomes zero. Thus, the force can be represented as,

$$A^o = KT^o_s(A^s - B)$$

where the diagonal matrix with three scale factors k_x, k_y, and k_z is represented by K and the column matrix with three biases: b_x, b_y, and b_z is depicted by B.)

7.2.2.2 Error model for gyroscope

Similar to accelerometer, an error model for Gyroscope can be formed as,

$\omega^o = KT_s^o(\omega^s - B)$, where ω^s is the angular velocity measured by gyroscope and ω^o is the gyroscope in orthogonal body coordinates.

7.3 Gait signal segmentation

A gait signal is cyclostationary in nature and generally is a sequence of multiple gait cycles. Thus, to model an individual's gait pattern, it is preferable to characterize each gait cycles instead of the complete signal. So, to extract gait cycles, one of the approaches is to find the periodicity of the signal. To find the periodicity of a signal, auto-correlation method, finding the dominant frequency using spectrum analysis, leveraging moving window are used. Another approach is to identify the point of beginning and the point of ending of each cycle. This is done by identifying one distinguishing event which is generally done by using zero-crossing method, especially for accelerometer. However, this may lead to uneven gait cycle lengths. For some machine learning techniques, it is required that the segments are of equal lengths. Interpolation and extrapolation can be applied on the extracted gait cycles in such cases.

7.4 Gait features using inertial sensor signals

A feature vector is used to define the characteristics of a gait pattern which can be used as a representation of a specific gait signal. Throughout the last decade, researchers have extracted different types of features of gait signals and studied their significance for different applications. In the next section, we will discuss about these features and how they can be derived from a sample signal.

7.4.1 Spatiotemporal features

These features embody either spatial or temporal information of a signal. Spatial features capture the change in space due to the movement, whereas temporal features represent time factors during the movement. Derivation of spatiotemporal features of gait requires identification of gait events such as heel strike, toe off, etc. In Chapter 3, the process of identifying these events has been discussed. The common spatial features that can be derived from inertial sensors in gait are stride length, step length, swing length, etc., and temporal features are cadence, single limb support, double support, etc., shown in Fig. 7.2. Brief descriptions to obtain these features are given below.

a) Stride length—This is defined as the distance covered by an individual's single stride, i.e., starting when one of the feet contacts the ground till the same foot again touches the ground. To extract stride length, one can measure the difference between two consecutive heel strike events in the signal and multiply with respective gait speed. This can be done with any of the other events also.

b) Step length—This is defined as the distance covered by an individual's single step, i.e., starting when one of the feet contacts the ground till the other foot touches the ground. To derive step length, the signals obtained from left and right foot need to be synchronized. The step length of left(/right) foot can be measured by multiplying gait speed with the time difference between one left(/right) foot heel strike and the next right(/left) foot heel strike.

c) Swing phase length—This is defined as the time required for an individual to complete swing phase during a single stride, i.e., the time period starting when one foot leaves the ground till the same feet touches the ground. For this, both toe-off and heel strike event in a gait cycle are to be identified. The difference between toe-off and next heel-strike event when multiplied with corresponding gait speed is the swing phase length.

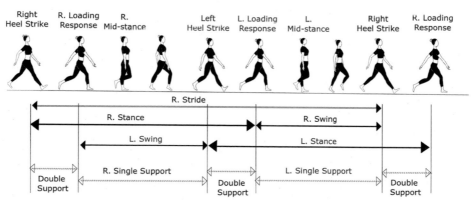

Figure 7.2 Different phases of gait cycle (presented for the right foot stride).

d) Stance phase length—This is defined as the time required for an individual's stance phase during a single stride, i.e., the time period starting when one foot touches the ground till the same feet leaves the ground. For this also, both toe-off and heel strike event in a gait cycle are to be identified. The difference between heel strike and next toe-off event when multiplied with corresponding gait speed is the swing phase length. It can also be derived by subtracting swing phase length from the stride length.

d) Cadence—This is the number of complete gait cycles covered by an individual in a minute. It depends on the speed of the individual's gait. This can be deduced by frequency of a gait event, e.g., heel strike, in an interval of specific number of timestamps in a minute depending on the sensor frequency.

e) Single limb support—This is defined by the period when only one foot supports the entire body during gait stance phase. This can be derived in a similar way to the stance phase length.

f) Double limb support—This is defined by the period when both feet are in contact with the ground and supports the body weight in a gait cycle. For this, the signals from both legs need to be synchronized and in addition to that, both toe-off and heel strike events in each gait cycle are to be identified. It can be derived from the time difference between the heel strike of one foot till the next toe-off event in another foot.

7.4.2 Statistical features

These features represent the statistical characteristics of a gait signal. Generally, these features are derived from each cycle/segments of a gait signal. Some of the examples of statistical features are mean, median, standard deviation, Kurtosis, skewness, etc.

7.4.3 Time-frequency distribution-based features

These features are derived from the time-frequency distribution of a gait signal. There exist multiple methods for time-frequency analysis of a signal such as fast fourier transform, short-time fourier transform, Wavelet transform, Wigner distribution, Hilbert-Huang transform, etc. These transforms generate the distribution of the signal with respect to time/space and frequency. Multiple features such as mean frequency, maximum/dominant frequency, skewness of the frequency values, approximate components, detailed components, etc., can be derived from these distributions.

7.4.4 Energy and entropy-based features

Energy can be measured of a finite signal by integrating squared amplitude values over some samples. Energy features such as spectral energy density can be useful, especially for analyzing disturbances. Similarly, entropy is also useful to characterize a signal. Different entropy measurements can be used such as Shannon's entropy, log energy entropy, etc.

7.5 Automated feature extraction using deep learning techniques

Hand-crafted features can be sometimes in-effective to represent an individual's gait pattern. It is also time-consuming to select and identify which features are more significant as it depends highly on the application. Owing to the advancement of deep learning techniques, feature extraction process can be automated and perfected with respect to target application. Deep neural networks, i.e., convolutional neural networks, auto-encoders can be used to analyze signals extract useful features to create a representation of the signal. This can help categorize as well as characterize different types of signals. By automating the feature extraction process, the deep learning framework has greatly affected the research domain of image and signal processing [33–38]. 1-dimensional convolutional neural network (1D-CNN) is generally used for the classification of one-dimensional (1D) signals. This architecture often consists of several kernels that produces output (also referred as tensor) by convolving with the given 1D spatial/temporal input signals [39–42]. However, most of the studies available in the literature related to wearable sensors used the architecture of the deep neural networks for human activity recognition [43–47]. Limited studies are available for analyzing the clinical gait pattern using deep learning utilizing wearable sensors. In [48], a classification based on convolutional neural network (CNN) is done by Ravi et al., In their experiment, they combined spectrogram and shallow features, i.e., mean, root mean square, variance, standard deviation of the original signal and its derivative, etc. A similar study is done by Dehzangi et al. in [49]. They used continuous wavelet transform algorithm to generate time frequency distribution images. Multiple sensors are used to acquire these images which is further used to train 2D-CNN as an input for classification. In [50], a deep convolutional neural network is trained by concatenating spectral information obtained from two consecutive windows/segments. Afterward, freezing of gait in Parkinson's disease patients is detected by testing the trained 1D-CNN model. Xia et al. proposed a method based on deep convolutional neural network which utilizes wearable accelerometer data to automate the feature learning process [51].

7.6 Gait pattern modeling using machine learning techniques

After obtaining well representative features, the next step is to select the most efficient machine learning technique to model different gait patterns. As mentioned before, some studies have used only statistical features, some frequency features, and some researchers have used different types of features. In this section, we will discuss only those methods which are more dominant for inertial sensor-based gait analysis irrespective of the type of features used.

Support vector machine (SVM)—SVM has advantage over other methods for classifying with high dimensional features. During gait analysis, the number of features may rise due to usage of multiple sensors and a large number of features can be generated. Also, another advantage of SVM is that the decision function can be modified with user specified kernels. In Ref. [22], the authors compared the performance of SVM with artificial neural network (ANN), Bayesian Belief Networks, and radial basis function network for classification of walking conditions in which SVM performed 100% accuracy. Mannini et al. used SVM to classify features extracted using group specific hidden Markov models (HMMs) [9]. Zhang et al. used SVM with radial basis function kernel to classify patterns between fatigue and nonfatigue gait characteristics with help of inertial sensor data [23]. For classification of elderly patients with chronic balance disorders, performance of different SVM models was satisfactory [24].

Tree-based classifier—Various tree-based classifiers are used for gait signal classification with satisfactory results. In (barth2012combined), the authors used Adaboost classifier to classify the signals obtained during both hand and gait motor functions of Parkinson's disease patient. Tripoliti et al. tested random forest (RF) models along with decision tree, random tree, naïve Bayes models to detect freezing of gait where RF reached highest accuracy of 96.11% [25]. In an experiment to classify the characteristics of imbalanced walking of Hemiplegic patients, authors used RF with forward search algorithm for feature selection and claimed to attain 100% accuracy [12].

Neural Network—Among different types of neural network architecture, the most commonly used for inertial sensor-based gait analysis is multilayer perceptron (MLP) networks. MLP has also been successfully used to classify the acceleration gait signals from patients with complex regional pain syndrome [13], also stroke patients and patients with neurological disorder (hsu2018-multiple). Other than MLP, ANNs have also proven to be equally good for pathological gait classification. Early researches such as in Ref. [14] showed the ability of ANN classifiers reaching 98% accuracy with on-board inertial sensors. Recent researches also find ANNs with various back-propagation algorithms can efficiently classify Parkinson's disease patients [15], CP children with spastic diplegia [16].

Apart from the techniques mentioned above, other popular classification techniques such as HMMs [17,18], fuzzy logic [19,20], k-nearest neighbor [21], clustering [26], etc., have also been used by some researchers to model gait patterns.

7.7 An example study

To give a brief idea about how inertial sensor data can be used for identification of pathological gait, we present a sample case study. Five healthy participants volunteered for this case study. The gait data are collected from each participant wearing the wearable sensor module around the shanks. The participants are instructed to walk on a treadmill normally for five times and again, five times by simulating Equinus gait. The speed of the treadmill is fixed at 3 kmph. A captured image during the data collection procedure is shown in Fig. 7.3. Written consents of each participant were taken before the experiment. The demographics of the participants are given in Table 7.1.

7.7.1 Experiment1: using handcrafted features

For this experiment, the gait data collected are first filtered to remove unwanted noise. Fig. 7.4 shows the signal obtained by accelerometer and gyroscope during walking. As mentioned in Chapter 5, signal segmentation is performed to extract gait cycles using zero-crossing, shown in Fig. 7.5.

The following features are obtained from the extracted gait cycles: (1) conventional features: stance time, swing time, heel strike amplitude; (2) statistical features: mean, median, variance,

Figure 7.3 A participant simulating Equinus gait while walking on a treadmill.

Table 7.1 Description of demographic information of the subjects of each candidate group.

Group	Count	Height (cm)	Weight (kg)	Age range (years)
Male	3	162.48 ± 12.2	62.5 ± 8.7	18—28
Female	2	155.37 ± 06.3	57.2 ± 7.2	18—27

maximum, minimum, kurtosis, skewness, root mean square, interquartile range, mean, and variance of first order derivatives; (3) frequency features: spectral density (Welch's method) with window length 12, and statistical features of the obtained spectral density. It is to be noted that here, stance/swing phase length is linear to stance time and swing time period as the speed of each individual is fixed at 3 kmph. The total number of features obtained considering the signals from both accelerometer and gyroscope signals is 194. The extracted features from the gait cycles are annotated with respective labels—0 for normal gait and 1 for Equinus gait. Next, for feature selection, importance of each feature is measured as the total reduction of Gini index brought by the specific features. This is done by fitting a RF classifier.

The feature importance values of 20 features with high feature importance values are shown in Fig. 7.6. Out of 194, finally top 100 features with high importance values are selected for training the classification models.

The set of 100 features along with class labels are divided in the following way: 20% is used for testing. The rest 80% of data

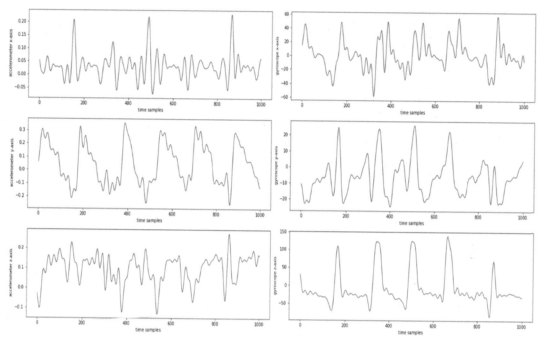

Figure 7.4 Accelerometer and gyroscope data obtained by Sparkfun inertial measurement unit sensor during walking.

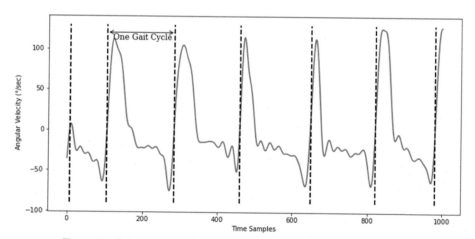

Figure 7.5 Gait cycles extracted from a gait signal using zero crossing.

is fivefold cross-validated during training of the models. Three different machine learning–based models: SVM, RF, and MLP are trained, validated, and tested. This is a binary classification problem, where class 0/normal is positive class, 1/Equinus is

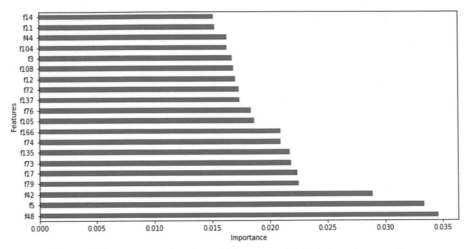

Figure 7.6 Importance plot of top 20 features with high importance values.

negative. The metrics used for evaluation are accuracy, sensitivity, specificity, positive predictive value, negative predictive value (NPV), and area under curve (AUC). Accuracy is defined by the ratio of correctly classified test samples out of total number of samples, whereas the specificity here denotes the number of samples classified as Equinus out of all Equinus gait samples. Detailed description about the classification evaluation metrics can be found in Ref. [27]. The validation results, i.e., average metric values of the results obtained for all folds are shown in Table 7.2. The test results, i.e., the results obtained on the unknown test dataset are shown in Table 7.3. The receiver operating characteristics (ROC) curve of each model during testing phase is shown in Fig. 7.7.

The results obtained during validation and testing show similar pattern. From the result, it is observed that RF has the

Table 7.2 Average of evaluation results of various machine learning method—based models for fivefolds during cross-validation process.

Method	Accuracy	Sensitivity	Specificity	PPV	NPV	AUC
SVM	66.19	61.67	81.96	92.18	48.29	0.75
MLP	60.87	62.13	59.93	53.54	68.05	0.65
RF	90.69	88.57	93.45	94.32	86.77	0.97

Table 7.3 Evaluation results of various machine learning method—based models during testing.

Method	Accuracy	Sensitivity	Specificity	PPV	NPV	AUC
SVM	64.91	60.01	77.5	87.23	43.05	0.72
MLP	56.28	60.30	53.75	45.13	68.27	0.66
RF	89.82	85.89	**94.57**	95.03	**84.72**	**0.96**

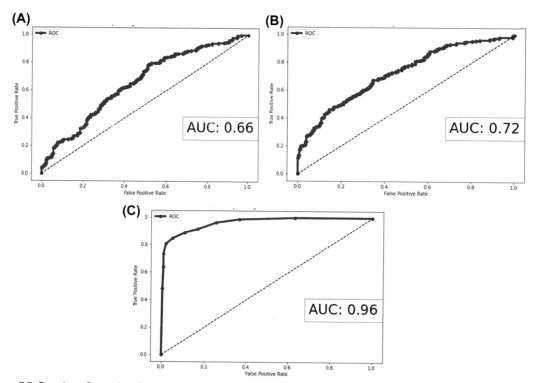

Figure 7.7 Receiver Operating Characteristics Curves of the performance of the models: (A) Multilayer Perceptron, (B) Support Vector Machine, and (C) Random forest.

best performance with highest accuracy score of ~90% during both training and testing. As our main objective is to identify the Equinus gait pattern, highest specificity and NPV values denotes the effectiveness of the RF model to detect Equinus gait pattern. Compared to MLP, SVM performed better obtaining 64.91% accuracy with unknown test data. However, in terms of NPV, MLP outperformed SVM with a difference of 25.22%. High

AUC value of ROC curve signifies that the RF model has higher separability between two classes than other models.

7.7.2 Experiment 2: using automated features

In this experiment, we considered deep learning techniques for extracting significant features instead of handcrafted features. The deep learning technique used for this study is: 1D-CNN. Preprocessing of the signals is done in similar way as in handcrafted features. However, the segmentation length is fixed to 150 samples instead of extracting cycles. This is done because 1D-CNN requires the input length to be of fixed dimension. The 1D-CNN architecture follows the architecture as described in Ref. [28], however with changed dimensions. Wavelet transform is applied to the signals obtained from three accelerometers and three gyroscopes of two IMU sensors. The decomposed signals are used as input vector.

The input vector has thus $(3 + 3) \times 2 = 12$ columns. The length of the input vector in the CNN model is 150 which is the segmentation length. Tuning the hyperparameters is done by changing the values of stride, size of filters, number of layers, activation functions, etc. Permutating the parameters, multiple times model is trained with different set of values and after a number of trials runs, the values which provided best results are selected as the finals parameter values. Table 7.4 describes the architecture is a multichannel deep neural network. It consists of five

Table 7.4 Filter size, output shape, and trainable parameters at each layer of the convolutional neural network architecture.

Layer	No. of filters	Filter size	Output shape	Parameters
Conv (ReLU)	64	3×1	(148, 64)	2368
Conv (ReLU)	64	3×1	146,64	12,352
Max-pooling	—	3	48,64	0
Conv (ReLU)	128	3	46,128	24,704
Drop-out	—	—	46,128	0
Conv (ReLU)	128	3	44,128	49,280
Drop-out	—	—	44,128	0
Conv	64	3×1	42,64	24,640
Global average pooling	—	—	64	0
Dense	—	—	2	130

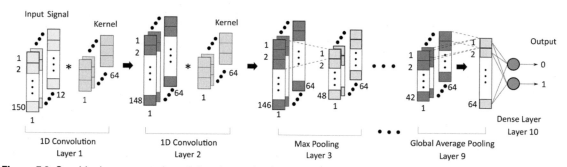

Figure 7.8 Graphical representation of the change in shapes of the output through different layers (Convolutional, Max Pooling, Global Average Pooling, Dense) of the proposed 1-dimensional convolutional neural network—based network.

convolutional layers followed by one max-polling layer. The architecture also includes dropout and average pooling layer and finally a dense layer.

Rectified linear unit (ReLU) is used as an activation function in each convolutional layer. From all available optimizers, we used "rmsprop" to compile the model and "binary cross entropy" as the loss function. Binary cross entropy is the most preferred loss function for binary classification as it fits the model with maximum likelihood function. Rmsprop is preferred as the optimizer because it is one of the fastest optimization algorithms. The "sigmoid" activation function is used for a fully connected dense layer as it is the most suitable for independent binary classes. The overall model has a total number of 113,474 trainable parameters in the framework. Fig. 7.8 presents the graphical representation of the change in shapes of the output through different layers.

Similar training and testing process is followed to evaluate system performance where 80% is used for training and 20% for testing. The result obtained for the 1D-CNN model is as follows: Train loss: 0.295, Train accuracy: 0.872, Testing loss: 0.559, Testing accuracy: 0.719. The accuracy and loss curve obtained during training and validation process are shown in Fig. 7.9. The precision, recall, and f1-score values obtained are 0.75, 0.68, and 0.71 for healthy class and 0.70, 0.76, and 0.73 for Equinus class. The ROC curve generated is shown in Fig. 7.10.

In the ROC plot, the AUC is 0.81, close to 1. It means that the proposed model gives a high degree of distinction between normal and Equinus gait pattern and can discriminate between the gait patterns of both classes' invariant to the classification threshold.

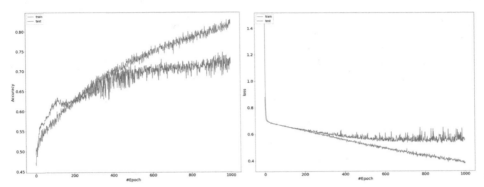

Figure 7.9 Change in accuracy and loss values with each epoch of training of the convolutional neural network model.

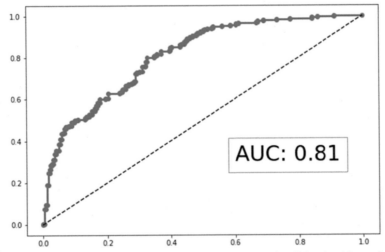

Figure 7.10 The receiver operating characteristics curve generated for classifying healthy and Equinus class with area under curve value 0.81.

7.8 Summary

In spite of improvement of inertial sensor technologies, inertial sensors are still not fit for complete assessment in clinical practice. Most of the existing studies mainly aimed at comparing pathological gait to healthy gait. However, it is important to make the system intelligent to differentiate between different pathological diseases as well as level of severity of the disease.

References

[1] M. Yoneyama, H. Mitoma, N. Sanjo, M. Higuma, H. Terashi, T. Yokota, "Ambulatory gait behavior in patients with dementia: a comparison with Parkinson's disease, IEEE Trans. Neural Syst. Rehabil. Eng. 24 (8) (2015) 817–826.

[2] E. Sejdić, K.A. Lowry, J. Bellanca, S. Perera, M.S. Redfern, J.S. Brach, Extraction of stride events from gait accelerometry during treadmill walking, IEEE J. Transl. Eng. Health Med. 4 (2015) 1–11.

[3] N. Hatanaka, et al., Comparative gait analysis in progressive supranuclear palsy and Parkinson's disease, Eur. Neurol. 75 (5–6) (2016) 282–289.

[4] B. Dijkstra, Y.P. Kamsma, W. Zijlstra, Detection of gait and postures using a miniaturized triaxial accelerometer-based system: accuracy in patients with mild to moderate Parkinson's disease, Arch. Phys. Med. Rehabil. 91 (8) (2010) 1272–1277.

[5] K.J. Kubota, J.A. Chen, M.A. Little, Machine learning for large-scale wearable sensor data in Parkinson's disease: concepts, promises, pitfalls, and futures, Mov. Disord. 31 (9) (2016) 1314–1326.

[6] J.C. van den Noort, V.A. Scholtes, J. Harlaar, Evaluation of clinical spasticity assessment in cerebral palsy using inertial sensors, Gait Posture 30 (2) (2009) 138–143.

[7] J.C. Van Den Noort, A. Ferrari, A.G. Cutti, J.G. Becher, J. Harlaar, Gait analysis in children with cerebral palsy via inertial and magnetic sensors, Med. Biol. Eng. Comput. 51 (4) (2013) 377–386.

[8] A.B. Bourgeois, B. Mariani, K. Aminian, P.Y. Zambelli, C.J. Newman, Spatio-temporal gait analysis in children with cerebral palsy using, foot-worn inertial sensors, Gait Posture 39 (1) (2014) 436–442.

[9] A. Mannini, D. Trojaniello, A. Cereatti, A.M. Sabatini, A machine learning framework for gait classification using inertial sensors: application to elderly, post-stroke and huntington's disease patients, Sensors 16 (1) (2016) 134.

[10] S. Yang, J.-T. Zhang, A.C. Novak, B. Brouwer, Q. Li, Estimation of spatio-temporal parameters for post-stroke hemiparetic gait using inertial sensors, Gait Posture 37 (3) (2013) 354–358.

[11] D. Trojaniello, A. Ravaschio, J.M. Hausdorff, A. Cereatti, Comparative assessment of different methods for the estimation of gait temporal parameters using a single inertial sensor: application to elderly, post-stroke, Parkinson's disease and Huntington's disease subjects, Gait Posture 42 (3) (2015) 310–316.

[12] J. Lee, S. Park, H. Shin, Detection of hemiplegic walking using a wearable inertia sensing device, Sensors 18 (6) (2018) 1736.

[13] M. Yang, H. Zheng, H. Wang, S. McClean, J. Hall, N. Harris, A machine learning approach to assessing gait patterns for complex regional pain syndrome, Med. Eng. Phys. 34 (6) (2012) 740–746.

[14] M.A. Hanson, H.C. Powell Jr., A.T. Barth, J. Lach, M. Brandt-Pearce, Neural network gait classification for on-body inertial sensors, in: 2009 Sixth International Workshop on Wearable and Implantable Body Sensor Networks, 2009, pp. 181–186.

[15] M.S. Baby, A.J. Saji, C.S. Kumar, Parkinsons disease classification using wavelet transform based feature extraction of gait data, in: 2017 International Conference on Circuit, Power and Computing Technologies (ICCPCT), 2017, pp. 1–6.

[16] Y. Zhang, Y. Ma, Application of supervised machine learning algorithms in the classification of sagittal gait patterns of cerebral palsy children with spastic diplegia, Comput. Biol. Med. 106 (2019) 33–39.

[17] M. Chen, B. Huang, Y. Xu, Human abnormal gait modeling via hidden Markov model, in: 2007 International Conference on Information Acquisition, 2007, pp. 517–522.

[18] C.M. Senanayake, S.M.N.A. Senanayake, Computational intelligent gait-phase detection system to identify pathological gait, IEEE Trans. Inf. Technol. Biomed. 14 (5) (2010) 1173–1179.

[19] M. Alaqtash, H. Yu, R. Brower, A. Abdelgawad, T. Sarkodie-Gyan, Application of wearable sensors for human gait analysis using fuzzy computational algorithm, Eng. Appl. Artif. Intell. 24 (6) (2011) 1018–1025.

[20] A. Laudanski, B. Brouwer, Q. Li, Activity classification in persons with stroke based on frequency features, Med. Eng. Phys. 37 (2) (2015) 180–186.

[21] J. Pauk, K. Minta-Bielecka, Gait patterns classification based on cluster and bicluster analysis, Biocybern. Biomed. Eng. 36 (2) (2016) 391–396.

[22] H.-Y. Lau, K.-Y. Tong, H. Zhu, Support vector machine for classification of walking conditions using miniature kinematic sensors, Med. Biol. Eng. Comput. 46 (6) (2008) 563–573.

[23] J. Zhang, T.E. Lockhart, R. Soangra, Classifying lower extremity muscle fatigue during walking using machine learning and inertial sensors, Ann. Biomed. Eng. 42 (3) (2014) 600–612.

[24] T. Nakano, et al., Gaits classification of normal vs. patients by wireless gait sensor and support vector machine (SVM) classifier, Int. J. Software Innovat. 5 (1) (2017) 17–29.

[25] E.E. Tripoliti, et al., Automatic detection of freezing of gait events in patients with Parkinson's disease, Comput. Methods Progr. Biomed. 110 (1) (2013) 12–26.

[26] Y. Sun, G.-Z. Yang, B. Lo, An artificial neural network framework for lower limb motion signal estimation with foot-mounted inertial sensors, in: 2018 IEEE 15th International Conference on Wearable and Implantable Body Sensor Networks (BSN), 2018, pp. 132–135.

[27] J.R.W. Morris, Accelerometry-a technique for the measurement of human body movements, J. Biomech. 6 (6) (1973) 729–736, https://doi.org/10.1016/0021-9290(73)90029-8.

[28] J. Chakraborty, A. Nandy, Discrete wavelet transform based data representation in deep neural network for gait abnormality detection, Biomed. Signal Process Contr. 62 (2020) 102076.

[29] R.B. Davis, Clinical gait analysis, IEEE Eng. Med. Biol. Mag. 7 (3) (1988) 35–40.

[30] A. Carriero, A. Zavatsky, J. Stebbins, T. Theologis, S.J. Shefelbine, Determination of gait patterns in children with spastic diplegic cerebral palsy using principal components, Gait Posture 29 (1) (2009) 71–75.

[31] A. Nieuwboer, R. Dom, W. De Weerdt, K. Desloovere, S. Fieuws, E. Broens-Kaucsik, Abnormalities of the spatiotemporal characteristics of gait at the onset of freezing in Parkinson's disease, Mov. Disord. Off. J. Mov. Disord. Soc. 16 (6) (2001) 1066–1075.

[32] A. Vienne, R.P. Barrois, S. Buffat, D. Ricard, P.-P. Vidal, Inertial sensors to assess gait quality in patients with neurological disorders: a systematic review of technical and analytical challenges, Front. Psychol. 8 (2017) 817.

[33] T. Wolf, M. Babaee, and G. Rigoll, "Multi-view gait recognition using 3D convolutional neural networks," in 2016 IEEE International Conference on Image Processing (ICIP), 2016, pp. 4165–4169, doi: 10.1109/ICIP.2016.7533144.

[34] A. Akula, A.K. Shah, R. Ghosh, Deep learning approach for human action recognition in infrared images, Cogn. Syst. Res. 50 (2018) 146–154, https://doi.org/10.1016/j.cogsys.2018.04.002.

[35] G. Altan, Y. Kutlu, A.Ö. Pekmezci, S. Nural, Deep learning with 3D-second order difference plot on respiratory sounds, Biomed. Signal Process. Control 45 (2018) 58–69, https://doi.org/10.1016/j.bspc.2018.05.014.

[36] R. Sharma, R. B. Pachori, and P. Sircar, "Automated emotion recognition based on higher order statistics and deep learning algorithm," Biomed. Signal Process. Control, vol. 58, p. 101867, 2020, doi: 10.1016/j.bspc.2020.101867.

[37] J. Werth, M. Radha, P. Andriessen, R. M. Aarts, and X. Long, "Deep learning approach for ECG-based automatic sleep state classification in preterm infants," Biomed. Signal Process. Control, vol. 56, p. 101663, 2020, doi: 10.1016/j.bspc.2019.101663.

[38] X. Zheng, W. Chen, M. Li, T. Zhang, Y. You, and Y. Jiang, "Decoding human brain activity with deep learning," Biomed. Signal Process. Control, vol. 56, p. 101730, 2020, doi: 10.1016/j.bspc.2019.101730.

[39] Y. Zheng, Q. Liu, E. Chen, Y. Ge, J.L. Zhao, Time series classification using multi-channels deep convolutional neural networks," in International Conference on We, Information Management (2014) 298–310, https://doi.org/10.1007/978-3-319-08010-9_33.

[40] B. Zhao, H. Lu, S. Chen, J. Liu, D. Wu, Convolutional neural networks for time series classification, J. Syst. Eng. Electron. 28 (1) (2017) 162–169, https://doi.org/10.21629/JSEE.2017.01.18.

[41] S. Bai, J. Z. Kolter, and V. Koltun, "An empirical evaluation of generic convolutional and recurrent networks for sequence modeling," arXiv Prepr. arXiv1803.01271, 2018.

[42] Y. Zhang, Y. Miyamori, S. Mikami, T. Saito, Vibration-based structural state identification by a 1-dimensional convolutional neural network, Comput. Civ. Infrastruct. Eng. 34 (9) (2019) 822–839, https://doi.org/10.1111/mice.12447.

[43] J. Yang, M. N. Nguyen, P. P. San, X. L. Li, and S. Krishnaswamy, "Deep convolutional neural networks on multichannel time series for human activity recognition," 2015.

[44] N. Y. Hammerla, S. Halloran, and T. Plötz, "Deep, convolutional, and recurrent models for human activity recognition using wearables," arXiv Prepr. arXiv1604.08880, 2016, doi: arXiv:1604.08880v1.

[45] M.A. Khan, K. Javed, S.A. Khan, T. Saba, U. Habib, J.A. Khan, A.A. Abbasi, Human action recognition using fusion of multiview and deep features: an application to video surveillance, Multimed. Tools Appl. (2020) 1–27, https://doi.org/10.1007/s11042-020-08806-9.

[46] M. Sharif, M. Attique, M.Z. Tahir, M. Yasmim, T. Saba, U.J. Tanik, A Machine Learning Method with Threshold Based Parallel Feature Fusion and Feature Selection for Automated Gait Recognition, J. Organ. End User Comput. 32 (2) (2020) 67–92, https://doi.org/10.4018/JOEUC.2020040104.

[47] H. Arshad, M.A. Khan, M. Sharif, M. Yasmin, M.Y. Javed, Multi-level features fusion and selection for human gait recognition: an optimized framework of Bayesian model and binomial distribution, Int. J. Mach. Learn. Cybern. 10 (12) (2019) 3601–3618, https://doi.org/10.1007/s13042-019-00947-0.

[48] D. Ravi, C. Wong, B. Lo, G.-Z. Yang, A deep learning approach to on-node sensor data analytics for mobile or wearable devices, IEEE J. Biomed. Heal. Informatics 21 (1) (2016) 56–64.

[49] O. Dehzangi, M. Taherisadr, R. ChangalVala, IMU-based gait recognition using convolutional neural networks and multi-sensor fusion, Sensors 17 (12) (2017) 2735.

[50] J. Camps, et al., Deep learning for freezing of gait detection in Parkinson's disease patients in their homes using a waist-worn inertial measurement unit, Knowledge-Based Syst. 139 (2018) 119–131.

[51] Y. Xia, J. Zhang, Q. Ye, N. Cheng, Y. Lu, D. Zhang, Evaluation of deep convolutional neural networks for detection of freezing of gait in Parkinson's disease patients, Biomed. Signal Process. Control 46 (2018) 221–230.

A low-cost electromyography (EMG) sensor-based gait activity analysis

8.1 Introduction

Wearable sensor-based gait analysis has great importance in clinical domain for diagnosis of different neuromusculoskeletal diseases. The advancement of wearable sensor-based technology provides a solution for cost-effective clinical gait analysis. During data analysis, the raw data can be acquired using wearable sensors from a person's body segment. The processed data are used for diagnostics, health care, or security purposes [1]. It is important to monitor step by step human's movement which filters a customer's natural signs, improvement precedents, or body present [1], and further performs development affirmation [2,3]. The subject needs to wear the sensor gadgets during data collection and also a subject feels comfort with minimum weight body worn sensors along with remotely data acquisition facility for movement-related applications [4,5]. It has been found in literature that Electromyography (EMG) signals have enormous applications in gesture recognition, controlling limb prosthetic, and gait analysis [6,7]. EMG signals offer muscle activation pattern for identification of different activities [8]. It has been observed that time domain (TD) features are mostly fast and easy to implement because no transformation is needed for these features [9]. Yeom et al. analyzed various TD features for finding the optimum way to reduce the complexity of implementation [10]. In frequency domain (FD) feature analysis, mean and median frequencies are used for feature extraction [11].

8.2 Description of lower leg muscles

It is understood that signification gait features can be obtained from lower extremity of human body. Therefore, the description

Modern Methods for Affordable Clinical Gait Analysis. https://doi.org/10.1016/B978-0-323-85245-6.00010-2

of lower leg muscles is important for clinical gait analysis. It can be described in the following manner:

(a) **Gastrocnemius**: It is a principle muscle located in the calves. The plantar flexion movement is realized in the ankle through this muscle. It results of pointing toe in downwards direction.

(b) **Soleus**: It is found that in back of the gastrocnemius, the large muscle is located to help with plantar flexion.

(c) **Plantaris**: It is also observed that in back of the lower leg, a small muscle is located which helps in plantar flexion like the gastrocnemius and soleus.

(d) **Tibialis**: The back and front portions of the lower leg contain these muscles. The muscles in the front allow for the dorsiflexion motion with pointing toes to upward. The back side muscles are responsible for plantar flexion along with the support of the arch of the foot.

(e) **Peroneus muscles**: These muscles are situated at the front portion of the lower leg which support with dorsiflexion.

8.2.1 What is gastrocnemius medialis?

The gastrocnemius muscle can be seen in lower leg muscles. It is composed of two strong locales, the average head and parallel head, which connect to the average and sidelong sides of the femur. The leaders of the gastrocnemius muscle cooperate to plantarflex the foot at the lower leg and to flex the leg at the knee.

8.3 Specification of MyoWare electromyography sensor

The electrical movements captured by skeletal muscles are preserved and accessed through EMG sensor. This sensor uses a device called an electromyograph to create an electromyogram. The EMG data are said to be proportionate with the amount of tension in the muscles. The muscle activities can be tracked directly through dynamic EMG. The intensity of muscle action can be processed through myoelectric signal which acts as an useful indicator for its mechanical effect. The relative muscle tension will be measured using amplitude of EMG signals obtained during gait acquisition. A number of surface EMG sensors are available in market for clinical applications. In our research, we are using MyoWare Muscle Sensor (AT-04-001) of Sparkfun cooperation which is very low-cost sensor with cost $40. Fig. 8.1 illustrates the MyoWare sensor layout.

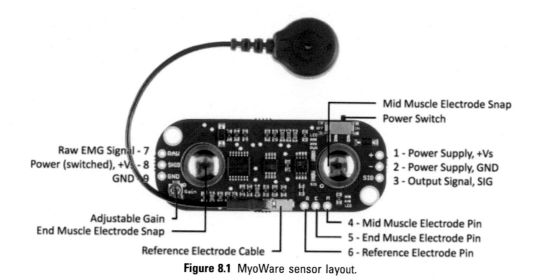

Raw EMG Signal - 7
Power (switched), +Vs - 8
GND - 9

Mid Muscle Electrode Snap
Power Switch

1 - Power Supply, +Vs
2 - Power Supply, GND
3 - Output Signal, SIG

Adjustable Gain
End Muscle Electrode Snap

4 - Mid Muscle Electrode Pin
5 - End Muscle Electrode Pin
6 - Reference Electrode Pin

Reference Electrode Cable

Figure 8.1 MyoWare sensor layout.

8.3.1 MyoWare Muscle Sensor (main board)

The most essential parts are three rightmost holes of the sensor. Having supplied voltage with + and − holes to the sensor, we collect the (rectified and integrated) EMG signals from the sensor from "SIG." It produces the analog output in the range of 0−1023. Fig. 8.2 depicts the specification of MyoWare sensor main board.

The sensor's top-leftmost side has a "RAW" signal output hole to collect the raw EMG signals at any point of situation. SHID'(+) and "GND"(−) can be used to power other electronic parts, if needed. Gain' controls how much amount is required to amplify the EMG signal. Fig. 8.3 also explains the MyoWare sensor connection modules in main board.

Figure 8.2 MyoWare sensor main board.

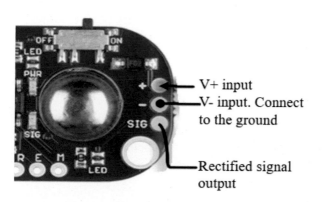

V+ input
V- input. Connect to the ground

Rectified signal output

Figure 8.3 MyoWare sensor main board.

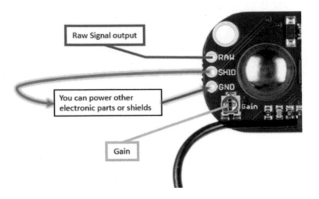

8.3.2 Myon electromyography sensor versus Sparkfun MyoWare electromyography sensor

8.3.2.1 Sparkfun MyoWare electromyography sensor

The muscle activity through the electric potential is measured by MyoWare which is commonly known as EMG. Our brain instructs our muscle to flex for receiving an electrical signal to our muscle for start recruiting motor units (the collections of muscle fibers which yield the force behind our muscles). The harder flex helps in recruiting more motor units to generate greater muscle force. The electrical activity of our muscle more increases with the larger number of motor units. The electrical activity will be analyzed by MyoWare sensor to produce an analog output signal which implies intensity of the muscles to be flexed.

The data were collected from healthy individuals including male and female. Before beginning the experiment, the subject's hair was removed to reduce the presence of noise, then the surface of the skin was cleaned with isopropyl alcohol/rubbing alcohol. The MyoWare EMG sensors were placed at gastrocnemius medialis muscle of the right leg. The reference electrode was connected to the right ankle as the reference is always connected to the bony region. Each person performed three gait-related activities: walking, running, and cycling each for a minute.

Fig. 8.4 depicts the placement of surface electromyography sensor where reference was connected to the ankle part.

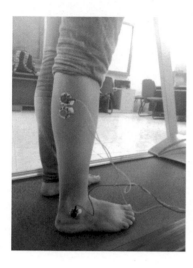

Figure 8.4 Surface electromyography placed at gastrocnemius medialis muscle of right leg.

8.3.2.2 Myon electromyography sensor

The Myon EMG sensor-based data collection was done in GV Lab, TUAT, Japan, with the different lab participants who contributed their gait patterns. In our experiment, we have used three channel Myon EMG sensors placed in different locations of lower extremity. The detailed description of data collection process is given in following Table 8.1.

The data were collected from five healthy subjects using surface EMG. These five subjects were made to walk on treadmill and over ground with various walking speeds and inclination. The data were collected from lower limb at 1000 Hz frequency. The electrical activity which is generated by skeletal muscles is required to evaluate and record through an electrodiagnostic technique known as EMG sensor. An electromyogram keeps a record using an instrument called as EMG after performing with EMG signal. The muscle activity is directly tracked by dynamic EMG. Fig. 8.5 shows the data collection in GV laboratory.

8.4 Hardware requirement for electromyography experimental setup

The MyoWare sensor by Advancer technologies is connected to Arduino pro mini. The Arduino pro mini 3.3 V/8 MHz is used

Table 8.1 Dataset description.

Participants list	Placement of sensors			Overground data description		Treadmill data description						
						Regular motion			Inclined motion			
									5%		10%	
	EMG_01	EMG_02	EMG_03	Slow walk	Fast walk	2 km	3 km	4 km	2 km	3 km	2 km	3 km
Subject1_ID0 H: 173 cm W: 61 kg	Right leg and left shank	Middle of thigh	Right leg and right shank	Multiple cycles for 60 s	Multiple cycles for 60 s	Multiple cycles for 120 s	Multiple cycles for 120 s	Multiple cycles for 120 s	Multiple cycles for 120 s	Multiple cycles for 120 s	Multiple cycles for 120 s	Multiple cycles for 120 s
Subject2_ID1 H: 177 cm W: 54 kg	Right shank front	Left shank front	Middle of thigh	Multiple cycles for 30 s	Multiple cycles for 30 s	Multiple cycles for 30 s	Multiple cycles for 30 s	Multiple cycles for 30 s	Multiple cycles for 30 s	Multiple cycles for 30 s	Multiple cycles for 30 s	Multiple cycles for 30 s
Subject3_ID2 H: 182 cm W: 65 kg	Right shank front	Left shank front	Middle of thigh	Multiple cycles for 30 s.	Multiple cycles for 30 s	Multiple cycles for 30 s	Multiple cycles for 30 s	Multiple cycles for 30 s	Multiple cycles for 30 s	Multiple cycles for 30 s	Multiple cycles for 30 s	Multiple cycles for 30 s
Subject4_ID3 H: 180 cm W: 65 kg	Right thigh	Left thigh	Right shank	Multiple cycles for 30 s	Multiple cycles for 30 s	Multiple cycles for 30 s	Multiple cycles for 30 s	Multiple cycles for 30 s	Multiple cycles for 30 s	Multiple cycles for 30 s	Multiple cycles for 30 s	Multiple cycles for 30 s
Subject5_ID4 H: 172 cm W: 77 kg	Right thigh back	Left thigh back	Left shank back	Multiple cycles for 30 s	Multiple cycles for 30 s	Multiple cycles for 30 s	Multiple cycles for 30 s	Multiple cycles for 30 s	Multiple cycles for 30 s	Multiple cycles for 30 s	Multiple cycles for 30 s	Multiple cycles for 30 s

Figure 8.5 Data collection in GV laboratory with different participants on mounting Myon surface electromyography sensors during walking on treadmill and over ground.

for the experiment. The raw signal pin of MyoWare sensor is connected to A0 of Arduino board. The power supply is provided by Lithium-ion battery of 3.7 V to the sensor through Arduino board. The positive pin of battery is connected to Raw pin of Arduino board and negative pin to the Gnd pin. MicroSD card adapter is used for data storage in the SD card. The adapter has a voltage regulator. Table 8.2 shows the pin connection between Arduino and MicroSD card adapter.

8.4.1 Arduino pro mini

The microcontroller board with Atmega328 is developed by Arduino.cc has right analog pins and 14 input/output digital pins. The size of this board is very small as compared to Arduino Uno (it is

Table 8.2 Pin Layout between MicroSD card adapter and Arduino pro mini.

MicroSD card adapter	Arduino pro mini
CS	10
SCK	13
MOSI	11
MISO	12
VCC	VCC
GND	GND

actually 1/6 of the total size of the Arduino Uno). The current version has one voltage regulator in the range of 3.3 V or 5 V and it runs at 8 MHz for the 3.3 V version. This pro mini board can be customized like other Arduino boards as per the requirement needs since the data and support-related facilities to this board are readily available. A flash memory of 32 KB is present in this board and out of which 0.5 is utilized for a bootloader. The use of flash memory is for storing the code of the board. It has nonvolatile memory which means if the connection with voltage supply is lost, the information will be still preserved. The Arduino pro mini board is illustrated in Fig. 8.6.

On the other hand, the 2 KB RAM memory is highly volatile, and it mainly depends on the constant source of power supply. It has an EEPROM with memory capacity of 1 KB. It is a read-only memory which is reprogrammable and can be erased using higher than normal electrical signals. The Arduino Software called integrated development environment is utilized to program the board. A sketch is used where the codes are written in board. Arduino Pro Mini has built-in LED like other available boards which will blink after certain programs get complied and run on the board. Fig. 8.7 shows the storage unit that is used with the pro mini board.

8.4.2 Interfacing with arduino pro mini

In existing work, sEMG is mostly used in analyzing or classifying upper-limb activities like hand gestures. For recognizing the lower limb activities, sensors like gyroscope or inertial motor unit are used. Since EMG gives the muscle activation pattern and every movement made is the result of muscle actions, so

Figure 8.6 Arduino Pro mini board.

Figure 8.7 MicroSD card adapter.

EMG sensors can be used for recognizing activities carried out using lower limb muscles.

The MyoWare sensor by Advancer technologies is connected to arduino pro mini.The Arduino pro mini 3.3 V/8 MHz is used for the experiment. The raw signal pin of MyoWare sensor is connected to A0 of Arduino board. The power supply is provided by Lithium-ion battery of 3.7 V to the sensor through Arduino board. The positive pin of battery is connected to Raw pin of Arduino board and negative pin to the Gnd pin. MicroSD card adapter is used for data storage in the SD card. The adapter has a voltage regulator. Table 8.2 shows the pin connection between Arduino and MicroSD card adapter. The interfacing with Arduino board is depicted in Fig. 8.8.

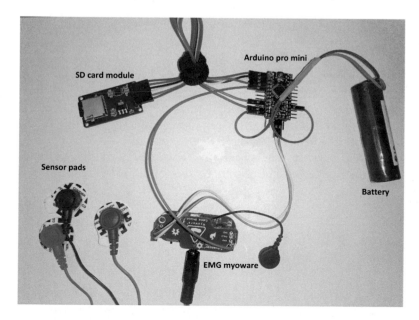

Figure 8.8 Interfacing with Arduino board for electromyography setup.

8.5 Preprocessing of electromyography signals

The EMG data capturers muscle activity pattern in the form of voltage. We stored the csv file which contains time versus voltage data. Dynamic EMG offers a means of directly tracking muscle activity. The myoelectric signal adequately parallels the power of muscle activity to fill in as a helpful marker of its mechanical effect. Amplitude of EMG signals derived during gait may also be interpreted as a measure of relative muscle tension. sEMG was used to get the muscle activation pattern of individuals while walking at different speeds and inclination on the treadmill. The raw EMG data were collected using sEMG and preprocessing and decomposition of the same was done to get the motor unit action potentials (MUAPs).

8.5.1 Filtering methods

Filtering is the processes of passing a signal through a filter circuit. The transfer function of a filter defines its behaviors. Filtering is used to eliminate noise from EMG signals and to smooth it. A low-pass butterworth filter is used here to smooth the signal. The cutoff frequencies of these filters have to be in the ranges of (5–30) Hz. Moreover, to prevent shifting of the signal in time, the filter should contain zero-phase delay properties. The raw and filtered EMG data are shown in Figs. 8.9 and 8.10, respectively.

8.5.2 Rectification

The EMG signal includes both positive and negative phases, which inconstancy about central line of zero voltage. Thus, it is presumed that the signal does not supply any useful information due to the fluctuation about zero value. Therefore, all negative amplitudes are rectified into positive amplitudes. This operation is called rectifying, done by taking the absolute value of all EMG data.

Figure 8.9 Raw electromyography data.

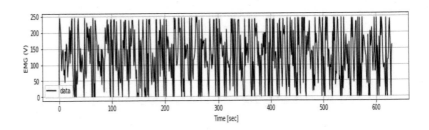

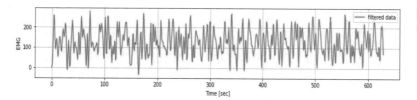

Figure 8.10 Filtered electromyography data.

8.5.3 Normalization

The peak dynamic normalization technique is used on the filtered EMG signal. Peak dynamic normalization takes the maximum value from the filtered data, and each value of filtered EMG data is divided by this maximum value to get the normalized signal.

From Figs. 8.11 and 8.12, it is observed that, when a person is walking with a speed of 2 km/h, the muscle activates only after first 1000 frames; on the other hand, if the person walks with a speed of 3 km/h, the activation starts nearly after 600 frames. The lag observed between two cycles for 2 km/h was around 1612 frames on an average, while 1236 frames for 3 km/h speed.

It has been observed from Figs. 8.13 and 8.14 that although the muscle activation starts at nearly same time, the lag observed is different for different speed. For 2 km/h, the lag between the cycles is nearly 1400 frames, and for 3 km/h speed, it is observed to be 1210 frames on an average.

8.5.3.1 Decomposition

EMG signals are the superposition of activities of multiple motor units. To reveal the mechanisms pertaining to muscle and nerve control, it is necessary to decompose the EMG signal. Different techniques have been devised with regards to EMG decomposition. The decomposition algorithm used consists of following processing stages: segmentation, PCA, and clustering. The advantage of this method is that it takes all frequency information in account and does not require manual selection of coefficients. The clustering technique used here is k means clustering. The signals are segmented, and on each segmented signal, PCA is applied which gives candidate MUAPs. K-means clustering is applied on these candidate MUAPs. The output of decomposition method is described in the following Fig. 8.15:

MUAP features like amplitude, phases, area, duration, and turns can be helpful to detect pathological issues in the person.

Figure 8.11 Walking on treadmill with 2 km/h.

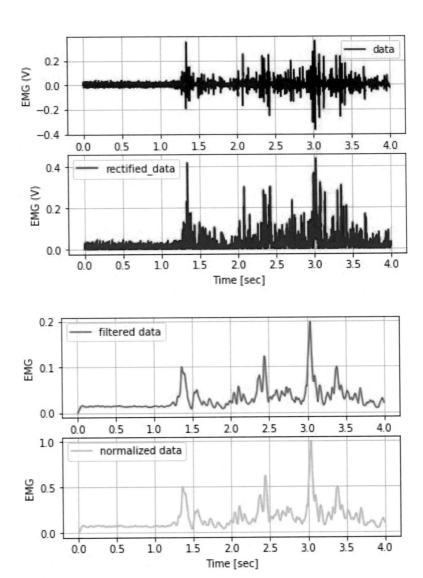

The MUAPs amplitude is observed to be above 0.8 V in most cases for a healthy person. Vital phases observed are three to four and turns observed are 4. EMG signal comes with the data to be appeared as the shape of linear envelope which mirrors the relative measure of muscle tension. It is understood that the relationship is impacted by strategy and physiological variables. Outlining the progressions in phasing, duration, or

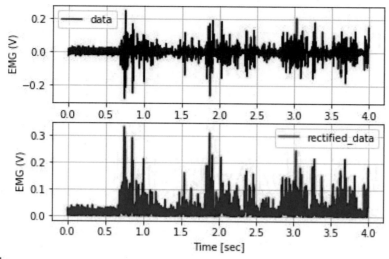

Figure 8.12 Walking on treadmill with 3 km/h.

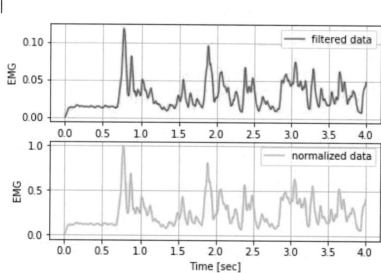

magnitude of muscle action related with a person's gait pattern, in any case, is troublesome because of electromyography record complexity. It's multispike, arbitrary amplitude quality challenges basic interpretation. The timing and intensity of the EMG during a gait cycle advises about neurological control and muscle integration.

Figure 8.13 Walking on treadmill with inclination (5%) with speed 2 km/h.

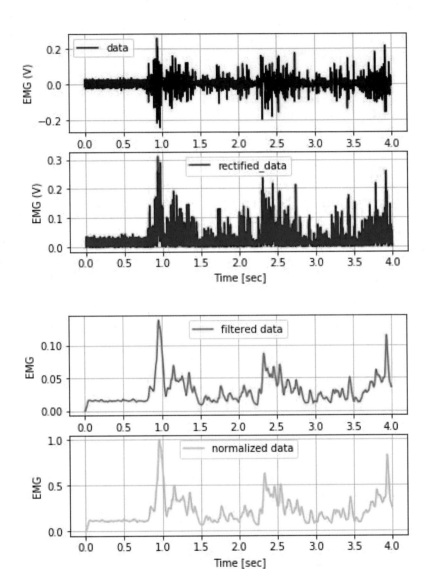

8.6 Electromyography sensor-based feature analysis

Feature analysis is an important step to extract intrinsic information from the processed data for classification of normal and abnormal gait pattern. We will explore two different types of feature extraction procedures for EMG-based gait analysis. The

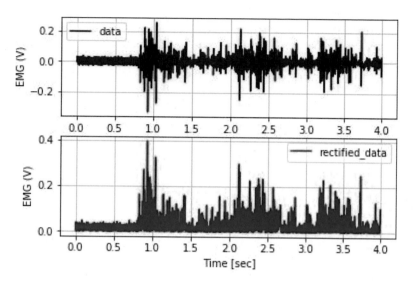

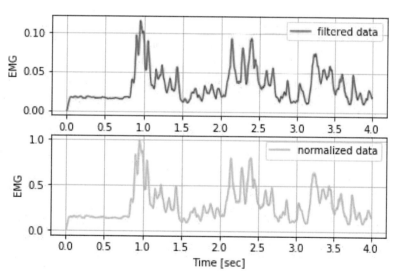

Figure 8.14 Walking on treadmill with inclination (5%) with speed 3 km/h.

detail descriptions of TD and FD features are given in the following way.

8.6.1 Time domain features

The TD features are simple and quick to be calculated on the time series signal itself, without any transformation [12]. In this

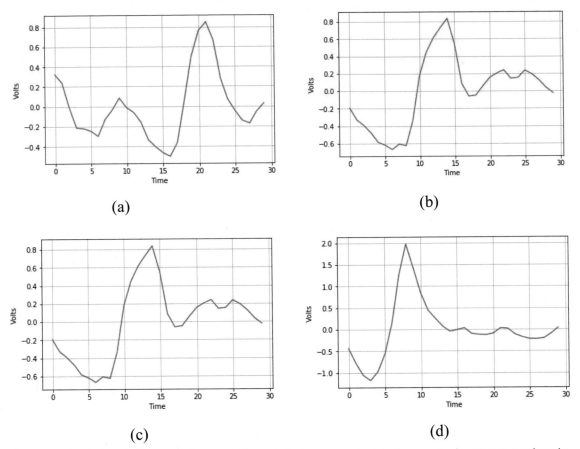

Figure 8.15 (A, B, C, and D) are the decompositions of electromyography into four most relevant motor unit action potentials.

study, we have used integral of absolute value (IAV), variance (VAR), and wavelength (WL) from various TD features.

8.6.1.1 Integral of absolute value

IAV is the absolute average value of each sample or segment, and it is calculated as follows:

$$\text{IAV} = \frac{\sum_{i=1}^{N} |S_i|}{N}$$

N = Number of samples.
S_i = signal value at instance i.

8.6.1.2 Variance

Variance is given by the average power of a random signal and can be explained as follows:

$$\text{VAR} = \frac{\sum_{i=1}^{N-1}(S_i - S)^2}{N}$$

8.6.1.3 Wavelength

It gives the length of the signal segment, which indicates the complexity of the signal:

$$\text{WL} = \frac{\sum_{i=1}^{N-1}|S_{i+1} - S_i|}{N}$$

8.6.1.4 Wilson amplitude

This is the count of the difference between two adjacent amplitudes exceeding a certain threshold. It can be formulated as:

$$\text{WAMP} = \sum_{1}^{N} u(|S_{i+1} - S_i| - T)$$

$$u(x) = \begin{cases} 1, & \text{if } x > 0 \\ 0, & \text{otherwise} \end{cases}$$

A threshold T of 0.05 V is considered in this analysis.

8.6.2 Frequency domain features

For FD features, the signal is first converted into the power spectrum density (PSD) [12]. In this analysis, we have used welch function to convert the signal in frequency spectrum. From FD features, mean frequency (MNF), peak frequency, and mean power (MNP) are used.

8.6.2.1 Mean frequency

The MNF of the EMG signal is formulated as the summation of product of the frequency and the EMG power spectrum, divided by total power:

$$MNF = \frac{\sum_{i=1}^{L}\left(P_j \times f_j\right)}{\sum_{i=1}^{L}\left(P_j\right)}$$

L = length of the frequency bin.
P_j = power at frequency bin j.
f_j = frequency of the spectrum at frequency bin j.

8.6.2.2 Peak frequency

The frequency at which the maximum peak occurs in power spectrum.

8.6.2.3 Mean power

The function given below gives the MNP of the spectrum.

$$MNP = \frac{\sum_{i=1}^{L}\left(P_j\right)}{L}$$

8.7 Gait analysis using surface electromyography sensors

The following classification models have been applied on the extracted feature for EMG sensor-based gait analysis.

8.7.1 Decision tree

Decision tree is a supervised learning technique with a tree-like structure. In this, all the internal nodes are some kind of test or an attribute and the terminal nodes usually represent the decision to be taken, i.e., class labels. Here we have used entropy metric for measuring the quality of split.

8.7.2 Random Forest

It is a type of ensemble classifier, where instead of using different algorithms, only decision tree is implemented. A set of decision tree is applied on different subsets of training set selected randomly with replacement. The aggregation of votes from different decision tree is taken into consideration for classification. In this experiment, we have ensembled 10 decision trees.

8.7.3 K nearest neighbor

It is a supervised algorithm, where an item is classified depending on the neighboring items. The class of an item is determined by calculating k nearest neighbor of the item, and then it is classified in to a class which is most common among those k neighbors. Here the value of k is used as five, and the metric used for calculating the distance is Euclidean distance.

8.7.4 Multilayer perceptron

It is also known as feed forward neural network. Multilayer perceptron (MLP) has minimum three layers. The first and last layers are input layer and output layer, whereas the middle layers are the hidden layers. The process is basically divided in three steps forward pass, calculating error and backward pass for learning and improving the weights. Here, we have used two hidden layers with 10 and 2 neurons. The activation function used is "relu," the rectified linear unit function, which returns $f(x) = \max(0, x)$.

8.7.5 XGBoost

It is an implementation of gradient boosted trees. It is a regularized model, which helps in improving overfitting, thereby increasing the performance accuracy. Here we have used 900 estimators as the accuracy after that became stagnant.

8.7.6 Support vector machine

Support vector machine (SVM) is a supervised learning technique, which can be used for both classification and regression. The objective of SVM is to give an optimal hyperplane that divides the data, which helps in classifying the new items. Here we have used penalty parameter C of the error term equal to 4.8.

8.7.7 State-of-the-art methods for electromyography sensor-based gait analysis

Different methods are needed before the feature extraction for managing EMG data to help increase accuracy. The data will be divided into segments, which has been through preprocessing techniques like filtering and normalizing. Different features will then be extracted from these segments and will be given as the input to the classifier. It has been pointed out by Englehart et al. [13] that the varying length of the EMG data has an effect on the classification error, which is also proven by Farina et al. [14]. Studies have proved that when the length of the segment was increased from 125 to 500 ms, the accuracy increases as well [15,16]. This might be because, larger the segment, more the information is obtained. This helps in getting a small bias and variance in feature estimation. To reduce the artifacts and noise in the EMG signals, it needs to be filtered. In Ref. [10], they have made a comparison on the basis of the performance of an adaptive filter. In Ref. [11], the authors have used a high pass and low pass filter with the cutoff frequency of 500 and 20Hz respectively. For leg muscles, Butterworth filter with varying cutoff frequency and orders is applied on the EMG signals. Table 8.3 shows different studies using different frequencies and order.

Feature extraction can be divided into three categories: TD, FD, and time-frequency domain (TFD) features. Hudgins et al. proposed five TD features: mean absolute value (MAV), zero crossing, mean value slope, WL, and slope sign change (SSC) [19]. In Ref. [20], MAV, SSC, VAR, and root mean square were used for detection of hand motions. Various features like WL and maximum amplitude (MAX) were then used with the other techniques. TD features are mostly fast and easy to implement because these features do not need any transformation [21]. Arief et al. analyzed various TD features for finding the optimum way to reduce the complexity of implementation [22]. Fuzzy approximate entropy (fApEn) and maximum voluntary contraction is used on the signals acquired from patients after stroke during their

Table 8.3 Various cut-off frequencies and orders used in the studies.

References	Cut-off frequency (Hz)	Order
[9]	6	4
[17]	5	6
[18]	5–500	6

rehabilitation training [23]. Few studies have used FD features in pattern recognition. FD features are used for recognizing muscle fatigue [24]. In Ref. [25], it is shown that the MNP frequency gives the knowledge about the changes in muscle signals in patients who survived stroke. For muscle contraction, features like PSD, MNF, and median power are preferred. Phinyomark et al. [21] modified the features to know the fatigue progression over time. Instead of calculating median and mean of power spectrum, they used amplitude spectrum. The modified versions were defined as modified MNF and modified median frequency. For detection of muscle fatigue, in Ref. [26], they used features like bandwidth, normalized spectral moments, MNF, and MDF. Comparison between the 27 TD features and 11 FD features is done by Ref. [27] for classifying hand movements. On the basis of the scatter plots, it was concluded that the TD features are redundant. Based on the output analysis and classifier, it is seen that though the TD features take less time to execute, the performance was not up to the mark [28]. The electrode pair selection is done using six TD features and five FD features by Ref. [17] where TD is found to be giving better performance than FD features for selection of electrodes. After extracting features from the EMG signals, the feature tuple is fed to a classifier. Authors in Ref. [29] expressed that the exhibition of highlight extraction and dimensionality decrease is reliant upon the abilities of a classifier. Factual classifiers, otherwise called LDA and MLP, had been utilized in the examination to characterize hand movements. The best execution is displayed utilizing LDA with a characterization exactness 93.75% when utilizing a PCA decreased list of capabilities. They additionally found that the MLP appreciates a bit of leeway over the LDA for being equipped for endorsing nonlinear class limits to include the abilities of the LDA. After 2 years, an LDA classifier performs well than an MLP classifier for the TFD-based highlights sets [30] as referenced in Ref. [29]. The LDA does not require heuristic determinations of its engineering or preparing calculation, yet it reliably performs great. This, probably, is because of the way that the PCA dimensionality decrease has an impact of linearizing the segregation undertaking of the classifier. LDA increased 98.87% of grouping exactness dependent on TD highlights [15]. Be that as it may, works done by Ref. [31] delivered about 99% order precision by utilizing MLP to arrange human lower arm movements dependent on TFD highlights. Ahsan et al. used ANN method for classifying hand gestures, where it gave 89.2% accuracy in one trail, when only TD features were taken into consideration [20]. Ref. [32] presents an automated gait mode identification from EMG signal of lower limb. In this approach, EMG signals were acquired at 1000 Hz sampling frequency with MyoTrace 400 (Noraxon, USA Inc.). This approach uses HMM for gait modeling with a modified version of Baum–Welch algorithm for training of the HMM parameters. It shows a good accuracy of classification. But it was pointed out by Ref. [33] that ANN takes time for training and it is not easy to decide the appropriate size of an ANN. SVM is a classifier which has given better performance has turned into a popular choice for machine learning techniques. It is difficult selecting the kernel function and other parameter values. The principal issues experienced in setting up the SVM model are the means by which to select kernel function and other parameters. In going for numerous clients to play out various movements, the bilinear model is proposed by creating two direct factors that are client based and movement based to recognize five hand signals utilizing SVM [16]. This technique brought about 73% precision, then was increased to 96.75% by hybridizing the particle swarm improvement and SVM [34]. These discoveries revealed that the kernel parameter setting of SVMs in classification of EMG signal is dependent on TFD features.

8.7.8 Classification models output with various gait activities

The entire process has been shown in Fig. 8.16. For classification of the three activities, performance comparison between six classifiers is done to get better classification accuracy. Metrics such as precision, recall, f1-score, and accuracy are used for the comparison. All the metrics are being calculated for individual activity.

Four TD features and three FD features are used for classifying three lower limb activities. The evaluation metrics discussed in Table 8.4 addresses that among all the six classifiers, XGBoost provides better performance with overall accuracy of 94.35%. The classifier which provides least performance is KNN, with the accuracy of 82.73%. In the existing studies, XGBoost has not been used for classification of activities based on EMG sensors. The comparison of classification algorithms based on feature domain is also illustrated in Table 8.5.

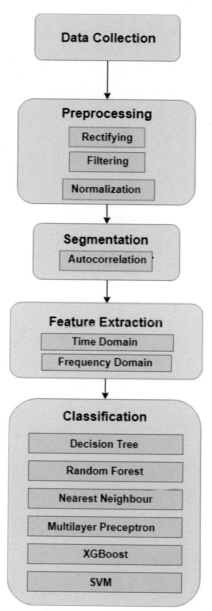

Figure 8.16 Work flow diagram.

Table 8.4 Comparison of different classifiers based on precision, recall, f1-score, and accuracy.

		Decision tree	Random forest	MLP	KNN	SVM	XGBoost
Precision	Walking	0.87	0.90	0.90	0.83	0.87	0.95
	Running	0.89	0.90	0.91	0.85	0.91	0.96
	Cycling	0.72	0.83	0.77	0.74	0.8	0.88
Recall	Walking	0.84	0.87	0.87	0.78	0.87	0.94
	Running	0.91	0.93	0.94	0.9	0.94	0.96
	Cycling	0.74	0.83	0.78	0.7	0.79	0.86
F1 score	Walking	0.86	0.89	0.89	0.8	0.87	0.94
	Running	0.90	0.92	0.92	0.87	0.92	0.95
	Cycling	0.73	0.83	0.76	0.72	0.74	0.87
Accuracy		0.86	0.89	0.89	0.82	0.88	0.94

Table 8.5 Comparison of classification algorithm based on feature domain.

Classifier	TD features	FD features	Combined features
Decision tree	71.72	80.9	86.48
MLP	84.2	85.4	89.7
Random forest	81.8	80.7	89.4
KNN	70.61	75.6	82.7
SVM	80.9	82.4	88.2
XGBoost	84.8	86.8	94.3

8.8 Summary

The skeletal muscles are responsible for any movement made by a human being. The activity carried out can be said to be the effect of activity in the muscles. The EMG signals give the muscle activation pattern during various motor action. The myoelectric signals show how the muscle actions have an impact on the mechanical movement. The amplitude of signals can be said to be proportionate to the relative muscle tension.

Different types of activities are identified using surface EMG signals in this analysis. The surface EMG data were collected from healthy individuals and the sensors were placed at Gastrocnemius Medialis muscle of the right leg. The sensor that we have used is MyoWare sensor by Advancer technologies. The MyoWare sensor is connected to Arduino pro mini for power and storage purpose. Each subject carried out three activities: walking, running, and cycling with the sensor placed.

The preprocessing techniques like rectification, filtering, and normalization were carried out on the acquired raw data before the signal was segmented into cycles using autocorrelation. Various time and FD features are extracted from each cycle. After feature extraction, different classifiers like random forest, extreme gradient boosting (XGBoost), MLP, and SVM are used for classification.

Comparison of the classifiers is done using metrics like total classification accuracy.

F-measure, precision, recall. Out of the six classifiers used in this study, XGBoost classifier achieves the best performance, with a total classification accuracy of 94.34%. The comparison between the TD and FD features showed that maximum classifiers give better performance with FD features. If the sEMG sensor is used along with other wearable sensors like inertial measurement unit sensors, then it may increase the classification accuracy. This analysis shows that the sensors such as EMG can be used for automated monitoring of activities. Apart from this, it also gives vital suggestions for other EMG signal-based devices, such as clinical applications, walking assist devices, and robotics or prosthetic gadgets.

References

[1] D.P. Tobón, T.H. Falk, M. Maier, Context awareness in WBANs: a survey on medical and non-medical applications, IEEE Wirel. Commun. 20 (4) (2013) 30–37.
[2] S.H. Roy, et al., A combined sEMG and accelerometer system for monitoring functional activity in stroke, IEEE Trans. Neural Syst. Rehabil. Eng. 17 (6) (2009) 585–594.
[3] J. Cheng, X. Chen, M. Shen, A framework for daily activity monitoring and fall detection based on surface electromyography and accelerometer signals, IEEE J. Biomed. Health Inf. 17 (1) (2013) 38–45, https://doi.org/10.1109/TITB.2012.2226905.
[4] S. Thongpanja, A. Phinyomark, P. Phukpattaranont, C. Limsakul, Mean and median frequency of EMG signal to determine muscle force based on time-dependent power spectrum, Elektron. ir Elektrotechnika 19 (3) (2013) 51–56.
[5] C. Zhu, W. Sheng, Wearable sensor-based hand gesture and daily activity recognition for robot-assisted living, IEEE Trans. Syst. Man Cybern. Part A Syst. Humans 41 (3) (2011) 569–573.

[6] A.A. Adewuyi, L.J. Hargrove, T.A. Kuiken, An analysis of intrinsic and extrinsic hand muscle EMG for improved pattern recognition control, IEEE Trans. Neural Syst. Rehabil. Eng. 24 (4) (2015) 485–494.

[7] I. Vujaklija, D. Farina, O.C. Aszmann, New developments in prosthetic arm systems, Orthop. Res. Rev. 8 (2016) 31.

[8] M.-S.S. Cheng, Monitoring Functional Motor Activities in Patients with Stroke, Boston University, 2005.

[9] A.A.B.A. Nadzri, S.A. Ahmad, M.H. Marhaban, H. Jaafar, Characterization of surface electromyography using time domain features for determining hand motion and stages of contraction, Australas. Phys. Eng. Sci. Med. 37 (1) (2014) 133–137.

[10] H. Yeom, ECG Artifact Removal from Surface EMG Using Adaptive Filter Algorithm, 2012.

[11] A. Balbinot, G. Favieiro, A neuro-fuzzy system for characterization of arm movements, Sensors 13 (2) (2013) 2613–2630.

[12] D. Tkach, H. Huang, T.A. Kuiken, Study of stability of time-domain features for electromyographic pattern recognition, J. NeuroEng. Rehabil. 7 (1) (2010) 21.

[13] K. Englehart, B. Hudgins, et al., A robust, real-time control scheme for multifunction myoelectric control, IEEE Trans. Biomed. Eng. 50 (7) (2003) 848–854.

[14] D. Farina, R. Merletti, Comparison of algorithms for estimation of EMG variables during voluntary isometric contractions, J. Electromyogr. Kinesiol. 10 (5) (2000) 337–349.

[15] A. Phinyomark, F. Quaine, S. Charbonnier, C. Serviere, F. Tarpin-Bernard, Y. Laurillau, EMG feature evaluation for improving myoelectric pattern recognition robustness, Expert Syst. Appl. 40 (12) (2013) 4832–4840.

[16] T. Matsubara, J. Morimoto, Bilinear modeling of EMG signals to extract user-independent features for multiuser myoelectric interface, IEEE Trans. Biomed. Eng. 60 (8) (2013) 2205–2213.

[17] C. Kendell, E.D. Lemaire, Y. Losier, A. Wilson, A. Chan, B. Hudgins, A novel approach to surface electromyography: an exploratory study of electrode-pair selection based on signal characteristics, J. NeuroEng. Rehabil. 9 (1) (2012) 24.

[18] H.M. Al-Angari, G. Kanitz, S. Tarantino, C. Cipriani, Distance and mutual information methods for EMG feature and channel subset selection for classification of hand movements, Biomed. Signal Process. Control 27 (2016) 24–31.

[19] B. Hudgins, P. Parker, R.N. Scott, A new strategy for multifunction myoelectric control, IEEE Trans. Biomed. Eng. 40 (1) (1993) 82–94.

[20] M.R. Ahsan, M.I. Ibrahimy, O.O. Khalifa, Neural network classifier for hand motion detection from EMG signal, in: 5th Kuala Lumpur International Conference on Biomedical Engineering 2011, 2011, pp. 536–541.

[21] A. Phinyomark, A novel feature extraction for robust EMG pattern recognition, J. Comput. 1 (1) (2009) 71–80. Available from: https://ci.nii.ac.jp/naid/10028227524/en/.

[22] Z. Arief, I.A. Sulistijono, R.A. Ardiansyah, Comparison of five time series EMG features extractions using Myo Armband, in: 2015 International Electronics Symposium (IES), 2015, pp. 11–14.

[23] R. Sun, R. Song, K. Tong, Complexity analysis of EMG signals for patients after stroke during robot-aided rehabilitation training using fuzzy approximate entropy, IEEE Trans. Neural Syst. Rehabil. Eng. 22 (5) (2013) 1013–1019.

[24] M.R. Al-Mulla, F. Sepulveda, M. Colley, A review of non-invasive techniques to detect and predict localised muscle fatigue, Sensors 11 (4) (2011) 3545–3594.

[25] X. Li, H. Shin, P. Zhou, X. Niu, J. Liu, W.Z. Rymer, Power spectral analysis of surface electromyography (EMG) at matched contraction levels of the first dorsal interosseous muscle in stroke survivors, Clin. Neurophysiol. 125 (5) (2014) 988–994.

[26] D.R. Rogers, D.T. MacIsaac, A comparison of EMG-based muscle fatigue assessments during dynamic contractions, J. Electromyogr. Kinesiol. 23 (5) (2013) 1004–1011.

[27] A. Phinyomark, P. Phukpattaranont, C. Limsakul, Feature reduction and selection for EMG signal classification, Expert Syst. Appl. 39 (8) (2012) 7420–7431, https://doi.org/10.1016/j.eswa.2012.01.102.

[28] A.-C. Tsai, T.-H. Hsieh, J.-J. Luh, T.-T. Lin, A comparison of upper-limb motion pattern recognition using EMG signals during dynamic and isometric muscle contractions, Biomed. Signal Process. Control 11 (2014) 17–26.

[29] K. Englehart, B. Hudgins, P.A. Parker, M. Stevenson, Classification of the myoelectric signal using time-frequency based representations, Med. Eng. Phys. 21 (6–7) (1999) 431–438.

[30] K. Englehart, B. Hudgins, P.A. Parker, et al., A wavelet-based continuous classification scheme for multifunction myoelectric control, IEEE Trans. Biomed. Eng. 48 (3) (2001) 302–311.

[31] R.N. Khushaba, A. Al-Jumaily, Fuzzy wavelet packet based feature extraction method for multifunction myoelectric control, Int. J. Biomed. Sci. 2 (2008) 126–134, 1.

[32] M. Meng, Z. Luo, Q. She, Y. Ma, Automatic recognition of gait mode from EMG signals of lower limb, in: 2010 The 2nd International Conference on Industrial Mechatronics and Automation, vol. 1, 2010, pp. 282–285.

[33] H.-B. Xie, T. Guo, S. Bai, S. Dokos, Hybrid soft computing systems for electromyographic signals analysis: a review, Biomed. Eng. Online 13 (1) (2014) 8.

[34] A. Subasi, Classification of EMG signals using PSO optimized SVM for diagnosis of neuromuscular disorders, Comput. Biol. Med. 43 (5) (2013) 576–586.

9

Low-cost systems—based therapeutic intervention

9.1 Introduction

Degraded gait and reduced dynamic stability deteriorate the quality of life. Proper therapeutic intervention can upgrade the gait quality of such patients. Subjects are asked to perform some specific tasks or activities during these interventions [1,2]. Sometimes they are given training [3] or asked to take specific drug [4] as a part of intervention. The effectiveness of an intervention is assessed by investigating the progress of patient. Recently, some inexpensive sensors like Kinect, inertial measurement unit (IMU), etc., are being used for such interventions. These sensors can also be used for monitoring patients to identify any need for change in the intervention process. When there is an urgent need of help, then a therapeutic intervention, i.e., a decision to take action can be applied to individuals to improve clinical support for the betterment of their life. The intervention includes psychological, physical, and pharmacological applied in different group of people in medical treatment, rehabilitation center. A gait assessment tool is applied to individual patient for measuring gait parameters and finding a solution for treatment plan to improve an individual's gait function. The diagnostic testing of gait parameters for children with cerebral palsy requires motion capture system synchronized with EMG system, force plate, etc. Miller et al., explores a study on children with ambulatory cerebral palsy for making treatment plan on pathological gait [5]. It is required to improve treatment plan of knee osteoarthritis (OA) through therapeutic options where OA is a painful disease affecting a maximum elderly population. It is an extremely complex disease which requires to study knee biomechanics during walking where kinematic and kinetic gait features are involved. The role of ambulatory biomechanics in OA is well understood through gait analysis for therapeutic interventions [6]. In the following sections, we discuss about the use of Kinect and wearable sensors for assisting in therapeutic interventions.

Modern Methods for Affordable Clinical Gait Analysis. https://doi.org/10.1016/B978-0-323-85245-6.00003-5

9.2 Kinect in therapeutic intervention

A few studies have used Kinect as a gaming device during intervention. Unobtrusive property of this sensor attracted the therapists. Sedef et al. [7] created virtual reality (VR) with the help of Kinect and assess its effect on motor functions of cerebral palsy population. The randomize control trial created two groups; one which attained the VR intervention and another one received the traditional occupational therapy. Evaluation was performed based on Bruininks-Oseretsky Test of Motor Proficiency-Short Form and assessed in accordance with independence in daily activities via the Wee Functional Independence Measure. After eight weeks of intervention, the VR group exhibited comparatively greater improvements. Garcia et al. [8] performed reaching of objects test for cerebral palsy patients. The author used shoulder Kinesio Taping on the patients for reaching-transporting of virtual objects. The environment was created using Kinect sensor (see Fig. 9.1). Schaham et al. [9] used Kinect-based video game for stroke rehabilitation. The authors reported Kinect as an efficient sensor for game-based interventions. Researches show that Kinect can be used for different therapeutic interventions.

Figure 9.1 Virtual environment created using Kinect.

9.3 Wearable sensors in therapeutic intervention

Apart from analysis of pathological gait pattern and severity of pathological conditions, wearable sensors also have been used for monitoring therapeutic interventions. For example, wearable sensors are used in monitoring elderly patients' movements for fall prediction; assisting clinicians for patients with Parkinson's disease by predicting freezing of gait (FoG); monitoring patients with musculo-skeletal injuries performing daily activities during rehabilitation; accelerometer is used for automated prosthetic limbs which can be operated with human feedback and assist amputees in performing daily activities, etc.

The injuries and fractures caused by falling of old age people are a critical issue. Obstacles in the home and aging are the most common cause of falls. Due to deterioration of vision, old age people cannot recognize tripping obstacle. Another reason which causes most falls occurs are the objects of things kept in the home. Usually, a home is filled with potential fall hazards. Low illumination or poor lighting condition, Slippery floors, unstable furniture, etc., are the common hazards inside a living environment. Machine learning techniques are proven to be an efficient method to detect fall detection and fall prevention. To overcome the issue of fall detection, many researchers produced many solutions. The two most common solutions are based on wearable sensors and the application of computer vision. A similar attempt is made by Ref. [10], where they used a wearable Shimmer device to develop an automatic fall detection system. The main objective of the proposed method is to reduce the size of data that is to be transmitted through a sensor to a computer via the wireless network.

Experiments are conducted on the database collected from 17 people, and three different systems are investigated and compared. Experimental results show an accuracy of up to 99.80% when a compressive sensing method is applied. A similar study based on the wearable sensor for fall related event detection is done by Bet et al. [11]. In their study, they tried to address the questions related to the type of sensors and their sampling rate, the type of signal and data processing employed, the scales and tests used in the study, and the type of application. A predictive model for fall detection based on the wearable sensor is investigated by Howcraft et al. [12]. Three different types of models, such as neural network, naïve Bayesian, and support vector machine, are investigated to classify fall risk. From the results, it is concluded that the neural network—based model performs best among all models. Compared to wearable sensor-based fall detection systems, camera-based fall detection systems use different approaches to detect falls. Some systems used human skeleton while other uses feature such as histogram projection, falling angle, etc. [13,14]. A detailed survey for fall detection based on both two approaches is done in Ref. [15]. However, there are few challenges associated with fall risk assessment, such as clinical fall risk assessment, sensor-based fall risk assessment, etc., and are discussed in Ref. [16].

Wearable sensors have been used for detection of FoG. Occurrence of FoG in a signal can be detected by binary classification techniques with distinguishing pattern between "non-FoG" and "FoG" class. Some studies present methodologies to predict the occurrence of FoG events early. These methodologies generally follow one of the following approaches. One approach suggests to create an extra class called "pre-FoG" and other approach uses time-series prediction algorithms and categorizes the predicted values to be of FoG or Non-FoG class. Mazilu et al. sliced sampled gait signals into partially overlapping windows. to 1s (64 samples) with 0:25s of overlap (16 samples) and used 60 statistical features for 3-class classification [17]. Palmerini et al. considered preFOG window length of 2s before the FoG episode [18]. Deep learning techniques are also implemented for the same purpose attaining higher performance [19]. Arami et al. implemented both above-mentioned prediction methods, i.e., 3-class classification and ARIMA model for predicting features and used those for prediction [20]. For the classification, SVM and probabilistic neural network are used. They correctly predicted FOG onset 77% for 3-class classification and around 85% for prediction-based classification recommending the later one for prediction of FoG. Daphnet is a widely used standard FoG datasets [21]. Wearable sensors are also used to monitor the progress with various tests that are performed on patients with neuro-muscular injuries during rehabilitation period. A detailed discussion is presented in chapter 10. Nowadays, the functional prosthetic limbs can be controlled in a variety of ways that help amputees to get back to normalcy. Advanced prosthetic legs can be equipped with a microprocessor and sensors that measure angles and forces while the patient walks. The measured values from the sensors can be used to make the microprocessor learns how the patient walks and constantly adapts the stiffness of the knee accordingly. Holmberg et al. devised an accelerometer-based autonomous control system for ankle prosthetic [22]. Kyberd et al. used accelerometer sensor values for the microprocessor controller to correct the amount of torque required to move the prosthetic limb [23]. Torrealba et al. introduced a bio-inspired and a feedback-based control strategy for the knee angle mechanism of the prosthesis [24]. sEMG sensors have also been used to improve the control systems of prosthetics [25—27].

9.4 Summary

Therapeutic interventions are the processes to help patients perform motor functions better. This chapter discusses about the applications of Kinect and wearable sensors in therapeutic interventions for pathological, elderly, or injured patients. We tried to highlight the usefulness of these sensors other than simple gait assessments.

References

[1] S. Khamis, R. Martikaro, S. Wientroub, Y. Hemo, S. Hayek, A functional electrical stimulation system improves knee control in crouch gait, J. Child. Orthop. 9 (2) (2015) 137—143, https://doi.org/10.1007/s11832-015-0651-2.

[2] Z.F. Lerner, D.L. Damiano, H.S. Park, A.J. Gravunder, T.C. Bulea, A robotic exoskeleton for treatment of crouch gait in children with cerebral palsy: design and initial application, IEEE Trans. Neural Syst. Rehabil. Eng. 25 (6) (2017) 650—659, https://doi.org/10.1109/TNSRE.2016.2595501.

[3] L.E. Mitchell, J. Ziviani, R.N. Boyd, A randomized controlled trial of web-based training to increase activity in children with cerebral palsy, Dev. Med. Child Neurol. 58 (7) (2016) 767—773, https://doi.org/10.1111/dmcn.13065.

[4] S. Galen, L. Wiggins, R. McWilliam, M. Granat, A combination of Botulinum Toxin A therapy and functional electrical stimulation in children with cerebral palsy - a pilot study, Technol. Health Care 20 (1) (2012) 1—9, https://doi.org/10.3233/THC-2011-0648.

[5] F. Miller, J. Henley, Diagnostic gait analysis use in the treatment protocol for cerebral palsy, in: B. Müller, S.I. Wolf, G.-P. Brueggemann, Z. Deng, A. McIntosh, F. Miller, W.S. Selbie (Eds.), Handbook of Human Motion, Springer International Publishing, Cham, 2017, pp. 1—15.

[6] J. Favre, B.M. Jolles, Gait analysis of patients with knee osteoarthritis highlights a pathological mechanical pathway and provides a basis for therapeutic interventions, EFORT Open Rev. 1 (10) (2016) 368—374.

[7] S. Aahin, B. Köse, O.T. Aran, Z. Bahadlr Ağce, H. Kaylhan, The effects of virtual reality on motor functions and daily life activities in unilateral spastic cerebral palsy: a single-blind randomized controlled trial, Game. Health J. 9 (1) (2020) 45—52, https://doi.org/10.1089/g4h.2019.0020.

[8] N. García-Hernández, J. Corona-Cortés, L. García-Fuentes, R.D. González-Santibañez, V. Parra-Vega, Biomechanical and functional effects of shoulder kinesio taping® on cerebral palsy children interacting with virtual objects, Comput. Methods Biomech. Biomed. Eng. 22 (6) (2019) 676—684, https://doi.org/10.1080/10255842.2019.1580361.

[9] N.G. Schaham, G. Zeilig, H. Weingarden, D. Rand, Game analysis and clinical use of the Xbox-Kinect for stroke rehabilitation, Int. J. Rehabil. Res. (2018), https://doi.org/10.1097/MRR.0000000000000302.

[10] O. Kerdjidj, N. Ramzan, K. Ghanem, A. Amira, F. Chouireb, Fall detection and human activity classification using wearable sensors and compressed sensing, J. Ambient Intell. Humaniz. Comput. 11 (1) (2020) 349–361.

[11] P. Bet, P.C. Castro, M.A. Ponti, Fall detection and fall risk assessment in older person using wearable sensors: a systematic review, Internet J. Med. Inf. 130 (2019) 103946.

[12] J. Howcroft, J. Kofman, E.D. Lemaire, Prospective fall-risk prediction models for older adults based on wearable sensors, IEEE Trans. Neural Syst. Rehabil. Eng. 25 (10) (2017) 1812–1820.

[13] Y. Hirata, A. Muraki, K. Kosuge, Motion control of intelligent passive-type walker for fall-prevention function based on estimation of user state, in: Proceedings 2006 IEEE International Conference on Robotics and Automation, 2006, pp. 3498–3503.

[14] A. Williams, D. Ganesan, A. Hanson, Aging in place: fall detection and localization in a distributed smart camera network, in: Proceedings of the 15th ACM International Conference on Multimedia, 2007, pp. 892–901.

[15] Y.S. Delahoz, M.A. Labrador, Survey on fall detection and fall prevention using wearable and external sensors, Sensors 14 (10) (2014) 19806–19842.

[16] T. Shany, S.J. Redmond, M. Marschollek, N.H. Lovell, Assessing fall risk using wearable sensors: a practical discussion, Z. Gerontol. Geriatr. 45 (8) (2012) 694–706.

[17] S. Mazilu, A. Calatroni, E. Gazit, D. Roggen, J.M. Hausdorff, G. Tröster, Feature learning for detection and prediction of freezing of gait in Parkinson's disease, in: International Workshop on Machine Learning and Data Mining in Pattern Recognition, 2013, pp. 144–158.

[18] L. Palmerini, L. Rocchi, S. Mazilu, E. Gazit, J.M. Hausdorff, L. Chiari, Identification of characteristic motor patterns preceding freezing of gait in Parkinson's disease using wearable sensors, Front. Neurol. 8 (2017) 394.

[19] V.G. Torvi, A. Bhattacharya, S. Chakraborty, Deep domain adaptation to predict freezing of gait in patients with Parkinson's disease, in: 2018 17th IEEE International Conference on Machine Learning and Applications (ICMLA), 2018, pp. 1001–1006.

[20] A. Arami, A. Poulakakis-Daktylidis, Y.F. Tai, E. Burdet, Prediction of gait freezing in Parkinsonian patients: a binary classification augmented with time series prediction, IEEE Trans. Neural Syst. Rehabil. Eng. 27 (9) (2019) 1909–1919.

[21] M. Bachlin, et al., Wearable assistant for Parkinson's disease patients with the freezing of gait symptom, IEEE Trans. Inf. Technol. Biomed. 14 (2) (2009) 436–446.

[22] W.S.U. Holmberg, An autonomous control system for a prosthetic foot ankle, IFAC Proc. 39 (16) (2006) 856–861.

[23] P.J. Kyberd, A. Poulton, Use of accelerometers in the control of practical prosthetic arms, IEEE Trans. Neural Syst. Rehabil. Eng. 25 (10) (2017) 1884–1891.

[24] A.F. Azocar, L.M. Mooney, L.J. Hargrove, E.J. Rouse, Design and characterization of an open-source robotic leg prosthesis, in: 2018 7th IEEE International Conference on Biomedical Robotics and Biomechatronics (Biorob), 2018, pp. 111–118.

[25] D. Hofmann, N. Jiang, I. Vujaklija, D. Farina, Bayesian filtering of surface EMG for accurate simultaneous and proportional prosthetic control, IEEE Trans. Neural Syst. Rehabil. Eng. 24 (12) (2015) 1333–1341.

[26] S.-K. Wu, G. Waycaster, X. Shen, Electromyography-based control of active above-knee prostheses, Contr. Eng. Pract. 19 (8) (2011) 875–882.

[27] B.H. Nakamura, M.E. Hahn, Myoelectric activation differences in transfemoral amputees during locomotor state transitions, Biomed. Eng. Appl. Basis Commun. 28 (06) (2016) 1650041.

Prevention, rehabilitation, monitoring, and recovery prediction for musculoskeletal injuries

10.1 Introduction

Musculoskeletal injuries are one of the most common injuries that can lead to long-term disabilities and may cause difficulty in even daily activities, and sometimes can be as fatal as to risk death. These injuries are frequent in occupations related to higher physical activities such as sports athletes, military personnel, construction workers, nurses, dancers, etc. Even heavy household chores also may lead to muscle injuries. Exercises like jumping, running cause injuries in the lower limbs, whereas throwing, lifting heavy objects cause injuries in the upper limbs and back. There is a misconception that injuries happen only because of carelessness or luck. However, if the causes of injury are thoroughly analyzed and proper prevention steps are taken, many injuries can be avoided. Thus, it is important to assess epidemiological factors of musculoskeletal injuries and carefully design preventive measures to minimize long-term disabilities. A preventive measure with cost-efficient sensors can be more beneficial in contrast to the difficulty of treatment of disability due to severe injuries. In this chapter, along with brief descriptions of the major musculoskeletal injuries in the lower limb muscles and about the various causes and treatments of these musculoskeletal injuries, we highlight the motion sensor-based solutions for prevention and recovery from injuries that are present in the literature. We focus on anterior crucial ligament injury which is a commonly occurring injury among sports athletes, especially for soccer players and have been frequently studied in literature.

Modern Methods for Affordable Clinical Gait Analysis. https://doi.org/10.1016/B978-0-323-85245-6.00004-7

10.2 Musculoskeletal injuries: causes and treatments

Among several types of musculoskeletal injuries, knee injuries are the most frequent that occur during sports activities [1]. Quadriceps, cruciate ligaments can be overstretched due to improper angle formation during landing or sudden turn in running. Lopes et al. reported that patellar tendinopathy has the highest incident rate among the marathon runners and plantar fasciitis being the most prevalent [2]. Achilles tendon muscle injury is also commonly reported among runners and soccer players. Occurrence of multiple Achilles tendinopathy can disable an individual for lifetime. Tendineal rupture of Hamstring muscle, if not taken care of properly, can be a long-term burden. The excessive loading and overuse of muscles during long haul physical activities has been the main stimulus for the development of the tendinopathies. Compared to knee injuries, ankle injuries are not considered fatal. Lateral ankle sprain can cause severe pain for long time. Plantar fasciitis is one of the foot injuries which is reported to be frequent during weight bearing. It can cause pain in the foot due to the excessive load on fascia. Although musculoskeletal injuries below the knee are more frequent, injuries on hip adductor have also comparable incidence rate. The causes and repercussions of the injuries of different parts of lower limbs are listed in Table 10.1 as mentioned by Kujala et al. [1]. Injuries on shoulder, forearms, thorax are not mentioned as those are out of scope of this book.

Musculoskeletal injuries are often treated with medications, therapeutic massage, or with aids like Kinesio taping, lumbar supports, etc. However, for severe injuries, surgeries such as soft tissue and cartilage repair, arthroscopy, etc., are needed to be performed depending on the type of injury. After surgery, patients have to go through a rehabilitation period to train themselves with simple exercises and tests and recover faster. Modification in lifestyle with the help of stretching, strengthening, and other exercises suggested by clinicians are recommended.

10.2.1 Anterior cruciate ligament injury

An injury that tears the anterior cruciate ligament (ACL) of the knee which causes severe pain restricting lower body movement activities such as walking, running, jumping, etc. People with ACL injury history are more prone to osteoarthritis (OA). In a study by Vaishya et. al, the authors conclude the rate of joint hypermobility was more common in patients with ACL injury than without

Table 10.1 Musculoskeletal injuries and their repercussions (of lower limbs) [1].

Region	Muscle	Injury	Repercussions
Thigh	Hamstring tendon	Complete rupture	Weakness, pain, and functional instability of the knee, chronic pain at the posterior thigh due to the lesions/neuromae of sciatic nerve branches
		Partial rupture	Pain at ischial tuberosity and posterior thigh at rest and particularly during activity, tightness of posterior thigh
Knee	Quadriceps ligament	Tear Contusion	Weakness, heterotopic ossification
	Cruciate ligament	Tear Injury	Instability, osteoarthritis (OA), limitations in daily activities
	Cartilage	Fracture	OA, stiffness
	Meniscus	Injury	Unilateral OA
Leg		Axial malposition	Varus/valgus of the knee and ankle, unilateral OA of the knee and ankle
	Epiphyseal		Growth disturbance
	Chronic compartment		Limitation and pain in physical activity
Ankle	—	Fracture	Decreased range of motion, OA
	—	Ligament instability	Functional instability, need to use brace, talo-tibial spur formation, limited dorsiflexion, OA
Foot	Tibialis posterior tendon	Tendon tear	Flatfoot and valgus deformity, mid-tarsal OA
	Navicular bone	Stress fracture	Pseudoarthrosis, talo-navicular OA
	Foot arc	Injuries/overuse	Flatfoot deformity, tarsal spurs
	First MTP joint	Injuries	Hallux rigidus
Pelvis and hip	Hip joint	Injury	Stiffness, OA
	Femoral collum	Stress fracture	Aseptic necrosis of caput femoris
	Iliac spine	Avulsion fractures	Spur, heterotopic ossification
	Ischial tuberosity	Avulsion fractures	Heterotopic ossification, nerve impingement

injury [3]. In investigations, the association of ACL injury with different sports, kabaddi and soccer constituted the highest percentages of ACL injuries whereas the highest association of lateral meniscal tear and chondral damage of the medial femoral condyle was found with kabaddi when compared with soccer [4,5].

ACL injuries are treated by replacing or mending torn ligament which is known as ACL reconstruction (ACLr). After the surgery, the patient goes through a training period to recover the injury and return to continue their physical (daily/sports) activities. John et al. reported that only around 39.8% of the athletes returned to sport and for those who were able to return to sports, 8.84 months is the average duration of time lost in sports [5].

10.3 Prevention of musculoskeletal injury through gait monitoring

The risk of injury increases with speed and violent contacts. As mentioned before, running, jumping, heavy lifting, and landing from certain heights can cause musculoskeletal injuries. In sports or military training, physicians are present to guide athletes on how to avoid injuries. However, it is not always possible for a human to monitor each and every athlete. Also, in scenarios like construction work or hospitals (for nurses) or at home, it becomes difficult to avoid injury. Yan et al. developed a real time motion warning personal protective equipment (PPE) for construction workers [6]. They use inertial measurement unit (IMU) sensor present in smartphones in the PPE. The IMU data for trunk inclination are processed in real time and checks if the angle is ergonomically safe. If the inclination crosses from "acceptable" zone to "not recommended," i.e., 60 degrees, the system alerts the subject until he fixes his posture to "acceptable" angle. Mauntel et al., devised a landing error scoring system for jump-landing assessment of US Military Academy cadets using Kinect depth camera [7]. The system autogenerates reports like asymmetrical foot contact, lateral trunk flexion, knee valgus, etc., that can be useful for clinicians. Mokaya et al. proposed an electromyogram (EMG) sensor-based wearable system, called Burnout, for skeletal muscle fatigue estimation which can be helpful for users against overuse of muscles [8]. Burnout obtains the region-based mean power frequency gradients which correlates the measured vibrations to Dimitrov's spectral fatigue index gradient, which is a known ground truth measurement of skeletal muscle fatigue. Myer et al. used force-plates during plyometric exercises for guiding the athletes on how to modulate force production during take off and force absorption during landing [9].

Kianifar et al. automated the single leg squat (SLS) test using IMU sensors [10]. SLS test a functional movement test that is carried out in rehabilitation, sports medicine, and orthopedic settings. This test is useful to measure the degree of inward movement of the knee, known as medial knee displacement or dynamic knee valgus (DKV). DKV is correlated with noncontact ACL injury and patellofemoral pain [12]. Dowling et al. developed a feedback-based training system based on kinematic measurements from inertial sensors to reduce ACL injury risk metrics (knee flexion angle, trunk lean, knee abduction moment) during jump landing [11]. Standardized movement modifications are suggested based on parameters such as knee flexion angle, trunk lean, and thigh coronal angular velocity. Seventeen subjects trained with instrumented feedbacks were able to reduce ACL injury risk metrics. Fig. 10.1 shows the setup of different sensors, i.e., IMU, force plates, EMG sensors used by researchers.

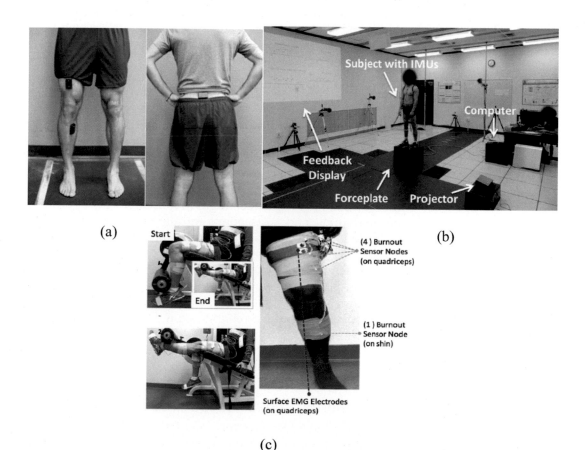

Figure 10.1 Experimental setup and placement of sensors of studies using different sensors. (A) Inertial measurement unit sensors [10] (B) Force plates and inertial measurement unit sensors [11] (C) electromyogram sensor [8].

10.4 Rehabilitation monitoring for recovery prediction

Patients who experience musculoskeletal injury require to go through the surgery that either replaces or mends the torn ligament. This surgery effects the normal movements of the patient and it requires around six months for ACL injury patients to fully recover [13]. During this recovering (ACLr) period, patients require to be regularly assessed and have to participate in various tests to monitor their recovery process such as hamstring and quadriceps strength testing, a negative pivot-shift test, hop testing: one-leg single hop for distance, the one-leg triple hop for distance, the one-leg timed hop, the one-leg crossover hop for distance, etc. The parameters such as laxity, isokinetic strength of hamstring and quadriceps are measured by clinicians during the tests as well as performing daily activities [14,15]. At the end of the recovery period, the pain should reduce as well as the strength difference between injured and noninjured should reduce increasing the limb-to-limb symmetry index. Muscle activation patterns alter following ACL injury and surgical repair. EMG sensor is commonly used wearable sensor to monitor muscle activation pattern. Changes in isometric muscle strength (MVC), voluntary activation, and surface electromyogram (sEMG) parameters are examined in relation to knee stability, pain, and swelling in patients. Drechsler et al. performed analysis on EMG median frequency and amplitude value which showed that there are subsequent changes in muscle fiber properties during detraining and subsequent retraining [16]. Mcginnis et al. examined leg motion and quadriceps muscle activation patterns during free-living activities [17]. They observed differences in muscle activation between affected and collateral limb within subjects and across subjects. In addition to that, differences in sEMG profiles, on the affected side, and gait kinematics are also observed with respect to rehabilitation duration (i.e., differences between 1 and 17 weeks post-op). The results prove that sEMG activation patterns and segmental kinematics during gait can be effectively used for rehabilitation.

10.4.1 Inertial sensor-based recovery prediction

Researches regarding recovery prediction using inertial sensors are still at preliminary level. There are only limited researches that focus on collecting IMU data with regular interval for monitoring purpose during ACLr period. Senanayake et al. developed an integrated motion analysis monitoring system by integrating three-dimensional (3D) kinematics and neuromuscular signals recorded using wearable motion and EMG sensors [18,19]. The data collected are first clustered into different groups for each activity using fuzzy clustering and fuzzy nearest neighbor methods. For each activity, separate segmentation is applied. Then features extracted from different activities are combined using Choquet integral fusion technique to predict recovery status of a subject. Authors of Ref. [20] investigated (1) the differences of spatiotemporal characteristics and sagittal plane shank angular velocity during loading response between limbs, (2) the correlation between the shank angular velocity during loading response of gait extracted from inertial sensors and the peak knee extensor moment calculated with a 3D motion capture system, (3) correlation between limb ratios of shank angular velocity and between limb ratios of knee extensor moment. The result analysis shows peak shank angular velocity and knee extensor moment were strongly positively correlated and the between limb ratios of angular velocity predicted 57% of the variance between limb ratios of knee extensor moment. In conclusion, authors suggest that inertial sensors have the potential to be an alternative to full motion analysis systems for identification of altered knee loading following ACLr. In Ref. [21], the authors pointed out that ACL injury does not alter spatial or temporal parameters of an individual's gait, but joint angular kinematics and kinetics with a very small degree of difference. They show that inertial sensor signal if used directly can extract more important gait characteristics such as rotation rate, rotation rate variance at different phases of cycle rather than deriving conventional spatiotemporal parameters such as gait cycle, stance and swing time, double support time, etc.

In Ref. [22] also, the authors state that classic measures of walking such as number of steps and gait speed, etc., are less informative than measurements of gait symmetry between injured and collateral foot for musculoskeletal injuries. However, they emphasized the feasibility of long-term digital monitoring of rehabilitation using waist-worn inertial sensor (accelerometer) and the potential to derive insights meaningful to both patients and practitioners. Kobsar et al. also suggested two sensor arrays at the back and thigh (thigh as the best position for sensor placement) may be the most ideal configuration to provide clinicians with an efficient and relatively unobtrusive way to optimize treatment [23]. However, Havens et al. in their experiment, found that Between-limb difference in thigh axial acceleration was the only variable which relates to between-limb difference in GRF, knee joint moments and power asymmetries [24]. The authors in conclusion stated that shank accelerations may not be as useful as a sensor placed on the thigh.

10.5 Summary

Musculoskeletal injuries can have major effect in an individual's health and career. Inability to work along with expensive treatments and rehabilitation procedures can in turn put the patients under economic stress. Monitoring the workers in high-risk jobs with inexpensive wearable sensors can prevent these situations. In this chapter, we thus focus on these inexpensive sensor-based preventive measures that have been conducted by researchers. We also discuss the research studies which have shown possible wearable sensor-based solutions to monitor injured patients during rehabilitation period. Rehabilitation centers can make use of these sensor-based monitoring to reduce human resource costs.

References

[1] U.M. Kujala, S. Orava, J. Parkkari, J. Kaprio, S. Sarna, Sports career-related musculoskeletal injuries, Sports Med. 33 (12) (2003) 869−875.
[2] A.D. Lopes, L.C. Hespanhol, S.S. Yeung, L.O.P. Costa, What are the main running-related musculoskeletal injuries? Sports Med. 42 (10) (2012) 891−905.
[3] R. Vaishya, R. Hasija, Joint hypermobility and anterior cruciate ligament injury, J. Orthop. Surg. 21 (2) (2013) 182−184.
[4] R. Gupta, T. Khanna, G.D. Masih, A. Malhotra, A. Kapoor, P. Kumar, Acute anterior cruciate ligament injuries in multisport elite players: demography, association, and pattern in different sports, J. Clin. Orthop. Trauma 7 (3) (2016) 187−192.
[5] R. John, M.S. Dhillon, K. Syam, S. Prabhakar, P. Behera, H. Singh, Epidemiological profile of sports-related knee injuries in northern India: an observational study at a tertiary care centre, J. Clin. Orthop. Trauma 7 (3) (2016) 207−211.
[6] X. Yan, H. Li, A.R. Li, H. Zhang, Wearable IMU-based real-time motion warning system for construction workers' musculoskeletal disorders prevention, Autom. Constr. 74 (2017) 2−11.
[7] T.C. Mauntel, et al., Automated quantification of the landing error scoring system with a markerless motion-capture system, J. Athl. Train. 52 (11) (2017) 1002−1009.
[8] F. Mokaya, R. Lucas, H.Y. Noh, P. Zhang, Burnout: a wearable system for unobtrusive skeletal muscle fatigue estimation, in: 2016 15th ACM/IEEE International Conference on Information Processing in Sensor Networks (IPSN), 2016, pp. 1−12.
[9] G.D. Myer, K.R. Ford, O.P. PALUMBO, T.E. Hewett, Neuromuscular training improves performance and lower-extremity biomechanics in female athletes, J. Strength Condit Res. 19 (1) (2005) 51−60.

[10] R. Kianifar, A. Lee, S. Raina, D. Kulić, Automated assessment of dynamic knee valgus and risk of knee injury during the single leg squat, IEEE J. Transl. Eng. Heal. Med. 5 (2017) 1–13.

[11] A.V. Dowling, J. Favre, T.P. Andriacchi, Inertial sensor-based feedback can reduce key risk metrics for anterior cruciate ligament injury during jump landings, Am. J. Sports Med. 40 (5) (2012) 1075–1083.

[12] D.A. Padua, D.R. Bell, M.A. Clark, Neuromuscular characteristics of individuals displaying excessive medial knee displacement, J. Athl. Train. 47 (5) (2012) 525–536.

[13] B.M. Cascio, L. Culp, A.J. Cosgarea, Return to play after anterior cruciate ligament reconstruction, Clin. Sports Med. 23 (3) (2004) 395–408.

[14] W. Micheo, L. Hernández, C. Seda, Evaluation, management, rehabilitation, and prevention of anterior cruciate ligament injury: current concepts, PM&R 2 (10) (2010) 935–944.

[15] D. Logerstedt, A. Lynch, M.J. Axe, L. Snyder-Mackler, Symmetry restoration and functional recovery before and after anterior cruciate ligament reconstruction, Knee Surg. Sport. Traumatol. Arthrosc. 21 (4) (2013) 859–868.

[16] W.I. Drechsler, M.C. Cramp, O.M. Scott, Changes in muscle strength and EMG median frequency after anterior cruciate ligament reconstruction, Eur. J. Appl. Physiol. 98 (6) (2006) 613–623.

[17] R.S. McGinnis, et al., Wearable sensors capture differences in muscle activity and gait patterns during daily activity in patients recovering from ACL reconstruction, in: 2018 IEEE 15th International Conference on Wearable and Implantable Body Sensor Networks (BSN), 2018, pp. 38–41.

[18] S.M.N.A. Senanayake, O.A. Malik, M. Iskandar, Wireless multi-sensor integration for ACL rehabilitation using biofeedback mechanism, in: ASME International Mechanical Engineering Congress and Exposition, vol. 45189, 2012, pp. 99–108.

[19] S.M.N.A. Senanayake, O.A. Malik, M. Iskandar, D. Zaheer, 3-D kinematics and neuromuscular signals' integration for post ACL reconstruction recovery assessment, in: 2013 35th Annual International Conference of the IEEE Engineering in Medicine and Biology Society (EMBC), 2013, pp. 7221–7225.

[20] S.M. Sigward, M.-S.M. Chan, P.E. Lin, Characterizing knee loading asymmetry in individuals following anterior cruciate ligament reconstruction using inertial sensors, Gait Posture 49 (2016) 114–119.

[21] M.R. Patterson, E. Delahunt, K.T. Sweeney, B. Caulfield, An ambulatory method of identifying anterior cruciate ligament reconstructed gait patterns, Sensors 14 (1) (2014) 887–899.

[22] A. Mueller, et al., Continuous monitoring of patient mobility for 18 months using inertial sensors following traumatic knee injury: a case study, Digit. Biomark. 2 (2) (2018) 79–89.

[23] D. Kobsar, S.T. Osis, J.E. Boyd, B.A. Hettinga, R. Ferber, Wearable sensors to predict improvement following an exercise intervention in patients with knee osteoarthritis, J. Neuroeng. Rehabil. 14 (1) (2017) 1–10.

[24] K.L. Havens, S.C. Cohen, K.A. Pratt, S.M. Sigward, Accelerations from wearable accelerometers reflect knee loading during running after anterior cruciate ligament reconstruction, Clin. Biomech. 58 (2018) 57–61.

Design and development of pathological gait assessment tools

11.1 Introduction

People with neuromusculoskeletal disorders have been looking for a reliable technology for automatic diagnosis of neurological and muscle abnormalities. There is an urgent need for pathological gait assessment tools in elderly age groups, sports person, army personnel who have been suffering for a long time from combating neuromusculoskeletal disorders. The research in this direction will certainly bridge the gap between engineering and biomedical field as it diagnoses and monitors of the neuromusculoskeletal abnormalities in an early stage which is both economically viable and convenient to be used on patients. An early detection technology allows us to take precautionary measure to delay the degeneration process.

An attempt has been taken to develop gait analysis tool with novel machine learning techniques for assessment of clinical gait patterns in biomechanical analysis system. This modern tool will be utilized to classify normal and abnormal gait. This clinical gait analyzer tool will have a great impact for practical applications in gait assessment of CP children. Furthermore, the developed pathological tool can be extended to apply on rehabilitation applications. These types of tools will ensure the quantitative measure for improvement of CP children after performing rehabilitation program. Early knowledge of the chances of fall will help them to be more cautious which will substantially decrease the events of injury. As a whole, this research will upgrade the quality of life of the CP children.

Modern Methods for Affordable Clinical Gait Analysis. https://doi.org/10.1016/B978-0-323-85245-6.00007-2

11.2 Tools for pathological gait assessment

11.2.1 Gait event annotation tools

Annotation of gait events of a signal obtained from nonstandard sensors are generally done by synchronizing with signals obtained from standard sensors. Pressure mats are the most standard devices to identify heel strike and toe off events. Although, there exist a number of methodologies for gait event detection, to the best of our knowledge, there exists one gait event annotation tool for multisensor gait signals [1]. The tool allows user to synchronize gait sensor signals such as gyroscope, accelerometer, or foot pressure sensor with corresponding video frame. The user can modify the labels by adjusting the synchronization. However, this tool has some drawbacks (1) for synchronization, recording of QR code using MotiCon software is required and as well as, jumps are required to be performed in the beginning and ending of the data collection procedure. This limits the application of this tool to any gait signal without this synchronization information. (2) If the vision sensor available has low-frame rate, the sensor signals are required to be down-sampled which may lead to loss of information. If no vision sensors available, then the annotation cannot be done. This limits the application of the tool only to the dataset having corresponding video stream from vision sensor with high frame rate. (3) The tool is developed as standalone application tool in Matlab, which is a proprietary software. This limits the access of the tool only to the users with access to Windows and Matlab. To overcome these limitations, we have developed a web-application tool, called VizGaitAssist (VGA) in our lab, which provides user assistance to annotate gait events in a gait signal without the requirement of synchronizing with any external video input. The web application tool is developed using open-source python language thus making it easily accessible and platform independent. The implementation and development details of VGA will be discussed in Section 11.3.

11.2.2 Gait diagnosis tools

Gait diagnosis tools are very useful for clinicians to assess gait by visualizing individual features. It also facilitates proper observation on progression of gait, performance of each trial, etc. Clinicians can perform manual annotation to extract joint or body parts specific features. In market, some available tools are there like ProtoKinetics Movement Analysis Software,[1] GAITSCAN,[2] RehaGait,[3] etc. These software provide a user-friendly environment to monitor and analyze gait. But, clinicians can only record data or extract features through these systems.

They have to perform further data processing to classify gait into targeted groups. Sometimes they have to conduct statistical tests on the extracted features or gait variables for decision making. In real sense, these systems do not automate the data analysis or classification. Again, those systems were made for costly sensors. This chapter demonstrates a graphical user interface (GUI)-based gait diagnostic tool which will automate the process of data analysis on the extracted features.

11.3 Development of a gait event annotation tool

We developed this application tool to help users who are looking for an easier way to annotate the gait events in a signal obtained from gyroscope signal or inertial measurement unit (IMU) sensor. The identification of gait events is required during the measurement of spatiotemporal features in a gait pattern. Although there exist gait event detection algorithms in the literature, sometimes highly quasi-periodic signals make it difficult to predict the events accurately. Specially, in case of slow speed gait, which are quasi-periodic in nature, the performance of the existing algorithms decreases. Therefore, user has to check and identify the events which are wrongly annotated. This process can become cumbersome. In our tool, we provide the users a visualization platform where they can easily rectify the errors with the help of interactive plots. Biomedical, biomechanical researchers, physiotherapists, and clinical gait analysis experts can use this tool for their convenience. To use this tool, it is required for the user to have rudimental knowledge about gait and gait events and also some basic knowledge to operate a computer system. It is a web-based application; no software package installation is needed for the user's system. As annotation is only a part of a complete gait analysis process, this application can be integrated with other gait analysis tools being used for clinical gait analysis.

---

[1]https://www.protokinetics.com/pkmas/.
[2]http://www.ohiinternational.com/products/technology/gaitscan/.
[3]https://hasomed.de/en/products/rehagait/.

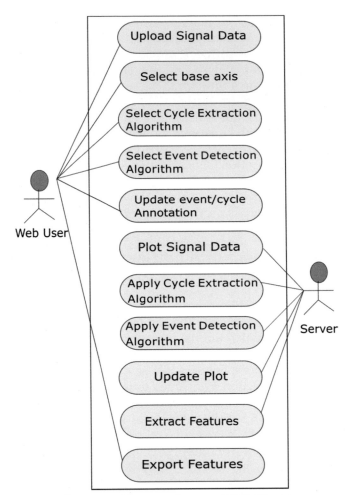

Figure 11.1 Use-case diagram of VizGaitAssist.

A use-case diagram of the developed tool is shown in Fig. 11.1. There are two actors interacting with the tool. First is the web user whose role is to upload the input signal and select different actions such as selecting the cycle extraction or the event detection algorithm to be performed on the signal. Server has the role to plot the signal and also update the plot with respect to the algorithm applied. Server will extract features from the annotated signal when the user commands to export the output.

11.3.1 User interface

Our "Visualization-based Assistive Tool for Clinical Human Gait Analysis" is basically a web-based application tool that consists

A Visualization-based Assistive Tool for Clinical Human Gait Analysis

This tool will help to pre-process gait signals and annotate gait events to extract spatio-temporal features

	IMU sensor	3-axis Gyroscope	1-axis Gyroscope
Module 1	Pre-processing	Pre-processing	Pre-processing
Module 2	Event Detection and Feature Extraction	Event Detection and Feature Extraction	Event Detection and Feature Extraction
Format	{accelerometer x-axis; accelerometer y-axis; accelerometer z-axis; gyroscope x-axis; gyroscope y-axis; gyroscope z-axis}	{gyroscope x-axis; gyroscope y-axis; gyroscope z-axis}	{ gyroscope axis}

NOTE: Only .csv files are supported as input files. Please make sure that the file does not contain any string values.

Developed by: Jayeeta Chakraborty and Dr. Anup Nandy, in Machine Intelligence and Bio-motion Research Lab, NIT Rourkela

Figure 11.2 Homepage of our web-based tool.

of mainly two modules: module1 is for preprocessing signals before feeding the signal for annotation process and module 2 is for annotation of gait signals. We have developed these two modules for three types of sensors: IMU sensors, 3-axis Gyroscope sensors, 1-axis Gyroscope sensors worn around shank of a person. The homepage of our tool contains the links to the two modules of the three types of sensors as shown in Fig. 11.2.

In module 1, the following options of preprocessing techniques can be applied to the obtained signal: (1) calibration, (2) clipping, (3) de-noising, and (4) filtering. A detailed description of the methods can be found in the "algorithms" section. The user interface (UI) for module 1 for IMU sensor is shown in Fig. 11.3. Each section of the interface of module 1 is described below.

Section 1.1: The "upload" button in the UI enables user to upload their data file into the tool (Fig. 11.4). The input file should be in ".csv" format only. The protocol for each type of sensor is mentioned in the homepage which should be followed.

Section 1.2: The simultaneous visualization of gait signals from the uploaded file can be observed in the interactive plots. These plots can be panned in and out, zoomed in and out according to requirement using the options available on the top of the plot. The option to reset to the original plot is also there. This

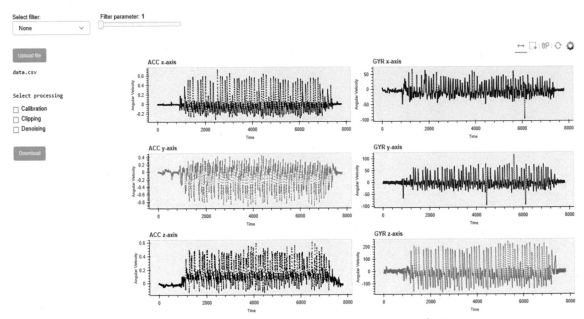

Figure 11.3 Interface of Module 1 for inertial measurement unit sensor.

allows the user to inspect the signal thoroughly both at microscopic and macroscopic level.

Section 1.3: The user can select different filtering techniques from the dropdown menu, "Select Filter" to be applied on the signal. He/she can also tune the filtering parameter of the signal according to their requirement using the slider, "Filtering parameter," and see the corresponding changes in the plots.

Section 1.4: In this section, the user can choose multiple techniques if it is required such as calibration, clipping, de-noising. Again, the effect of each of the operations is visible in the plots whenever selected.

Section 1.5: After applying the preprocessing techniques, if the user is satisfied with the end result, the final improvised data can be downloaded and saved in a ".csv" file. This downloaded file can be later used in the second module of our tool as it is saved in the required specific format.

Module 2 is our main module where the annotation of gait events can be done. The gait events that we have considered are: heel strike, foot flat, toe off, and mid swing. In this module, a user can annotate these events and also annotate the end of each gait cycle. Different types of spatiotemporal features can be obtained from the annotation of these events such as the length between two consecutive heel strikes is the cycle length

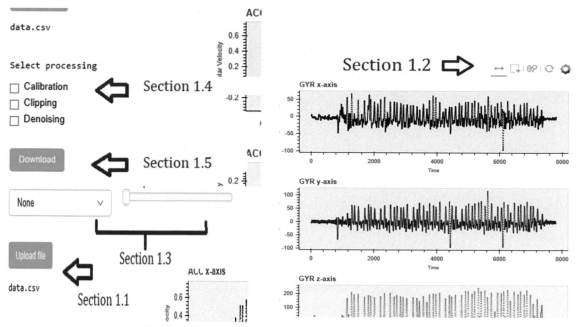

Figure 11.4 A closer look at different sections of module 1.

or the length between heel strike and toe off in a single cycle corresponds to the stance length of that cycle. The UI for module 2 for IMU sensor is shown in Fig. 11.5. Each section of the interface of module 2 is described below.

Section 2.1: Similar to module 1, the "upload" button in the UI enables user to upload their data file into the tool (Fig. 11.6). The

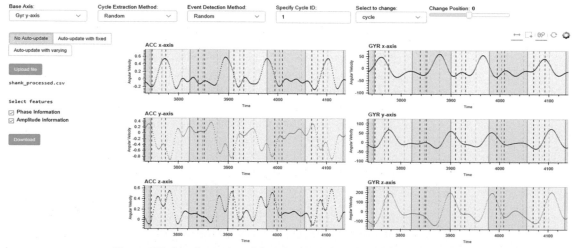

Figure 11.5 Interface of module 2 for inertial measurement unit sensor.

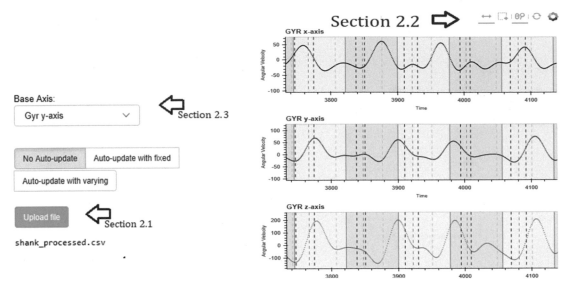

Figure 11.6 A closer look at Section 2.1, 2.2, and 2.3.

input file should be in ".csv" format only. The protocol for each type of sensor is mentioned in the homepage which should be followed. We recommend to use the file obtained from the module 1 to be used for module 2.

Section 2.2: The simultaneous visualization of gait signals from the uploaded file can be observed in the interactive plots. These plots can be panned in and out, zoomed in and out according to requirement using the options available on the top of the plot. The option to reset to the original plot is also there. This allows the user to inspect the signal thoroughly both at microscopic and macroscopic level.

Section 2.3: The first drop-down menu is the "base axis." When IMU data are captured, depending on the placement of the sensor, angular velocity and linear acceleration is measured along the *x*-axis, *y*-axis, and *z*-axis. If the user placed the sensor such that the gait movement is along the *y*-axis during their experiment, then user should select *y*-axis as "base axis"(Fig. 11.6). Basically, depending on this axis data, the cycle extraction and other methods will take place, so it is important to carefully select the base axis.

Section 2.4: The next option is to select a cycle extraction method from the drop-down menu, "Cycle Extraction Method" which consists of the following methods: "Random," "Autocorrelation Analysis," "Spectrum Analysis," "Event-based Extraction" (Fig. 11.7). Selecting "Random" option divides the signal into multiple cycles of a random length. The rest of the options

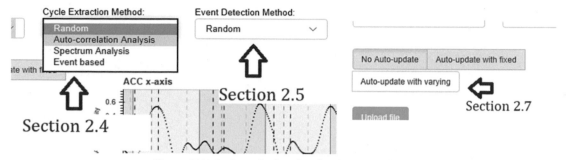

Figure 11.7 A closer look at Section 2.4, 2.5 and 2.7.

are the most popular methods that exist in the literature to identify gait cycles from a gait signal and will be discussed in a later section "Description of the Methods used."

Section 2.5: The section, "Gait Event Detection" is to detect and annotate events such as toe-off, mid-swing, heel-strike, foot-flat in the signal. Options in the "Gait event detection" methods change according to the selection of "Cycle Extraction Method" mentioned in *Section 2.4.* If "Event based" is selected, different machine learning techniques such as "Decision Tree," "Random Forest," and "XGboost" based models which are already trained are incorporated in this section. If "Event based" is not selected, then a generic method, "Maxima-minima" and optional "Random" may be applied to identify the events.

Section 2.6: The width of a cycle and its corresponding gait events can be modified using "Specify cycle id," "Select to change," and "Change Position" Fig. 11.8. If user wants to make any modification to a cycle, he/she has to first "Specify cycle id" then select which event he/she wants to change from the dropdown, "Select to change"; the options available are "cycle length," "Heel Strike," "Foot Flat," "Mid Swing," and "Toe Off"; then the user can change the position of the event in that cycle using the slider and see the corresponding changes in all the plots. Simultaneous visualization will help the user to make a decision even if one of the sensor values is erratic.

Section 2.7: An option to auto-update the signal with fixed cycle length and varying cycle length is integrated in our module 2 (Fig. 11.7). There is also option for "no auto-update," in case the user does not want cycles to be updated automatically. For example, when the user is changing gait cycle 4 and auto-update with fixed length option is active, the remaining cycles will be updated with the same width as in cycle 4. If auto-update with varying cycle width is active, then the remaining cycles will be extracted based on the shape of the current gait cycle.

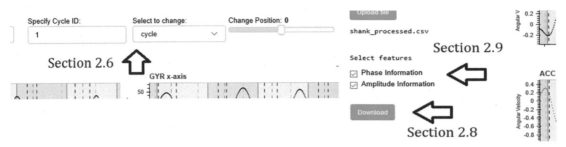

Figure 11.8 A closer look at Section 2.6, 2.8 and 2.9.

Section 2.8: The extracted features can be exported in a.csv file using the "Download" button. The output format of each sensor is different and is mentioned in the tool itself. For example, the output format of the spatio-temporal features for IMU sensor is <'cycle_id', 'cycle_len', 'swing_len', 'stance_len', 'toe_off_-amp_accx', 'toe_off_amp_accy','toe_off_amp_accz','toe_off_amp_gyrx','toe_off_amp_gyry','toe_off_amp_gyrz','mid_swing_amp_accx', 'mid_swing_amp_accy','mid_swing_amp_accz','mid_swing_amp_-gyrx','mid_swing_amp_gyry','mid_swing_amp_gyrz','heel_strike_amp_accx','heel_strike_amp_accy','heel_strike_amp_accz','heel_strike_amp_gyrx','heel_strike_amp_gyry','heel_strike_amp_gyrz', 'foot_flat_amp_accx','foot_flat_amp_accy','foot_flat_amp_accz', 'foot_flat_amp_gyrx','foot_flat_amp_gyry', 'foot_flat_amp_gyrz'>

Section 2.9: There is also provision for the user to export either phase information or amplitude information as per the requirement.

11.3.2 Description of the methods used

A brief description of the preprocessing methods used in module 1 is mentioned below.

(1) Calibration: Sensors due to structural error measures some signal even when the person is not moving. This leads to a difference in the measured output and expected output. The process to remove this error is called calibration. For calibration, first average of the measured value of the sensor is taken while the person is in standing position; this value is then subtracted from the rest of the measured values.

(2) Clipping: Before starting to walk or run, the data obtained when the person is not in motion are discarded. The process to clip these data are called clipping. It is done by tracking the minimum movement in the sensor and clipping it.

(3) De-noising: This process removes the signal noise that falls way outside the general measurements of the sensor.

(4) Filtering: This process generally removes the components with unwanted frequency from the signal. The user can specify up to what extent he/she may want to filter the signal by managing a threshold value. There are multiple filtering methods. We have implemented spectrum analysis based, butter-worth filtering for the filtering process of the gait signals.

A brief description of the cycle extraction methods is mentioned below.

(1) Auto-correlation analysis: A correlation value of a time series to itself is called auto-correlation. In auto-correlation analysis, correlation value with increasing lag is computed and nearest peak to the zero crossing value is considered to be the time period of the signal.

(2) Spectrum analysis: In this process, spectrum distribution of the signal against frequency is measured and the largest peak is considered to be the dominant frequency. The time period corresponding to this frequency is taken as the underlying periodicity of the signal.

(3) Event-based extraction: To detect the periodicity of a gait signal, sometimes gait events, generally heel strike, is taken as the start and end of the cycle. For this, we have to first detect heel strike event in a gait signal.

A brief description of the event detection methods is mentioned below.

(1) Minima-maxima: In this method, from a complete gait cycle different gait events are detected by finding the local minima and local maxima. Generally, the lowest valley point in the cycle is heel strike, and the second lowest valley point is toe off; the highest peak is mid-swing, and the second highest peak is foot flat. Although, due to presence of noise in wearable sensors, local maxima and minima is not always in par with the events.

(2) Machine learning based models: We have used some machine learning models and trained them to identify a gait event from a gait signal. We used gait data of 10 people walking in different speed from 1.5 to 4 km/h. First, we manually annotated each event of the gait signals. From the location of the events, we extracted different features; for example, for IMU sensor the features selected were: x-axis accelerometer and gyroscope value, y-axis accelerometer and gyroscope value, z-axis accelerometer and gyroscope value and the derivatives of each of these points, also previously occurred event of the

current event and distance between these two events. These set of features are divided into 80% for training and 20% for testing. A decision tree model, a XGBoost model and a random forest model are trained with the training data and tested on the test data. These models are used in our tool for the identification of gait events.

The auto-update techniques followed in the tool are given below.

(1) Auto-update with fixed length: This method will update the next cycle lengths based on the cycle length that user is currently specified in the section "Specify Cycle Id." It will not update the previous cycles assuming that the user has already gone through those cycles and made the required changes.

(2) Auto-update with varying length [2]: In this method, a sample cycle is first taken as the base cycle which is the current cycle selected in the current tool, then depending on this cycle the rest of the signal is divided into cycles of varying length. A window of (0.5*previous cycle length, 1.8*previous cycle length) is taken as range to find the next cycle assuming that two consecutive cycle lengths will not differ much. The cycle with maximum similarity to it's previous cycle within this range is selected as the next cycle. This method uses Kolmorogrov–Smirnov tests and Hausdorff's distance to measure the similarity between two cycles.

11.4 Development of a gait diagnosis tool

Existing automatic gait diagnostic tools use expensive motion detection sensors (e.g., Vicon, Qualisys cameras, etc.) which make it unaffordable for most of the clinics. To assist the clinicians, a novel low-cost gait detection tool has been developed through visual understanding of gait signal acquired from multiple Kinect v2 sensors. In this tool, there are two major functions which are preprocessing of raw input data and classification using different machine learning algorithms. Preprocessing includes interpolation, smoothing, transformation, trajectory merging, and outlier detection and removal techniques. The machine learning algorithms used for classification are long short term memory (LSTM), artificial neural network (ANN), and support vector machine (SVM). This tool provides a GUI-based platform which could enhance the visual interpretability of the outcomes. Biomedical, biomechanical researchers, physiotherapists, clinical gait analysis experts can use this tool for their convenience.

It is a standalone application and therefore the users must install the tool before using it. No other add-ons are required to run this application. In addition, being a low-cost system, this application carries high potential for clinics and rehabilitation centers.

11.4.1 Description of the data acquisition system

Data were collected from a multi-Kinect system (see Fig. 11.9). Fifteen children and adolescent with cerebral palsy were taken as abnormal group and age matched normal children (15) were taken as normal group. The Kinects were placed on tripods (Slik F153) sequentially at 35 degrees angle with the walking direction. The arrows emerging from Kinects cover the horizontal field of view of Kinect, i.e., 70 degrees. The distance of the Kinects from the left border of the track was 2 m. Height and tilt angle of the Kinects were 0.8 m and 0 degrees, respectively, while the distance between the Kinects was 3.5 m. The width of the track was 0.84 m. This setup allowed ~0.5 m of overlapped tracking volumes (between two successive sensors) which was empirically estimated to be sufficient (for both the groups) for the next sensor to recognize a person's body and start tracking. A computer was set as the server which controlled each of the Kinect connected to separate computers. System clocks of the computers were synchronized using Greyware's Domain Time II (Greyware Automation Products, Inc.), which follows PTP protocol. A training session was provided to the participants before the experiment.

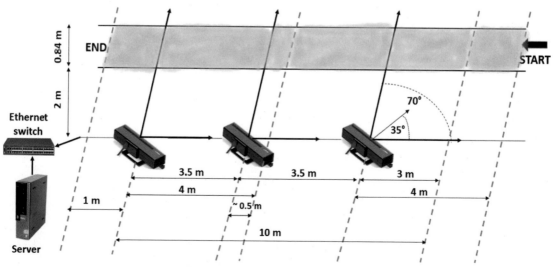

Figure 11.9 Data collection setup [3].

Subjects were asked to start walk at self-selected speed from 4 m distance from the first Kinect and after walking 1 m distance (marked by line) data were started to capture. Subjects were asked to walk up to the "End" line. The total distance of the path was 12 m out of which 10 m distance was considered for data collection. The extra distances (at start and ending points) were given to reduce the effect of acceleration and deceleration on gait variables. Five trials for each participant were taken with 2 min of resting gap. Clinicians have to follow the described protocol as a guideline to construct such acquisition system.

11.4.2 Development of the system

This tool is basically a standalone application software and its function is to detect the abnormality of human gait. Three-dimensional positional value of lower limb joints (i.e., ankle, knee, hip, etc.) were considered as raw input data. The software contains two major modules: (1) preprocessing the raw input signals and conversion the positional data to gait velocity and (2) detection of gait abnormality using different machine learning classifiers. The main page of our tool is shown in Fig. 11.10. The

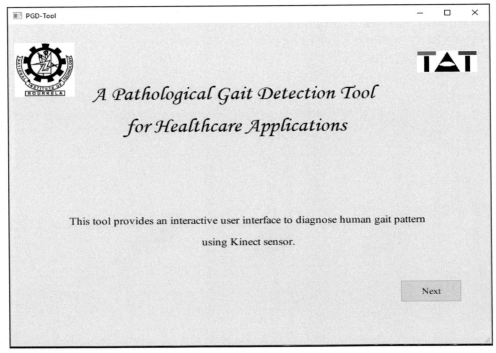

Figure 11.10 Main page of the tool.

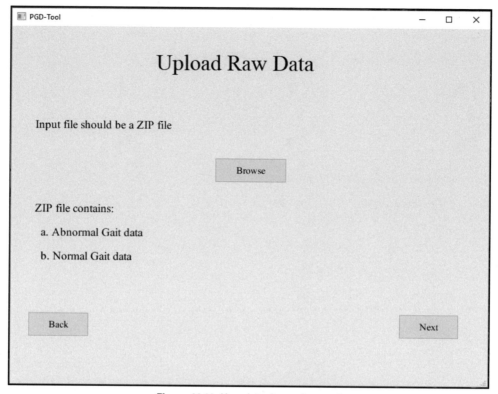

Figure 11.11 User Interface of page 2

name of the application is: *A Pathological Gait Detection Tool for Healthcare Applications* abbreviated as PGD.

The preprocessing technique applied to the signal includes: (1) interpolation, (2) smoothing, (3) transformation, (4) trajectory merging, and (5) outlier detection and removal. A detailed description of each process is given in the subsequent sections. The UI for page 2 of this tool is shown in Fig. 11.11.

The "Browse" button in the GUI enables the users to upload their data file into the tool. The input should be in ".zip" format and the files inside the *.zip file must contain only ".csv" files. The "Next" button is inactive until the uploading is not complete. Once the uploading is complete a pop-up message will be displayed informing the completion of the task.

Once the uploading is complete the user can click on the "Next" button to view the next page of the tool. The UI for page 3 of the tool is shown in Fig. 11.14. Preprocessing of raw data is performed in page 3 (Fig. 11.12).

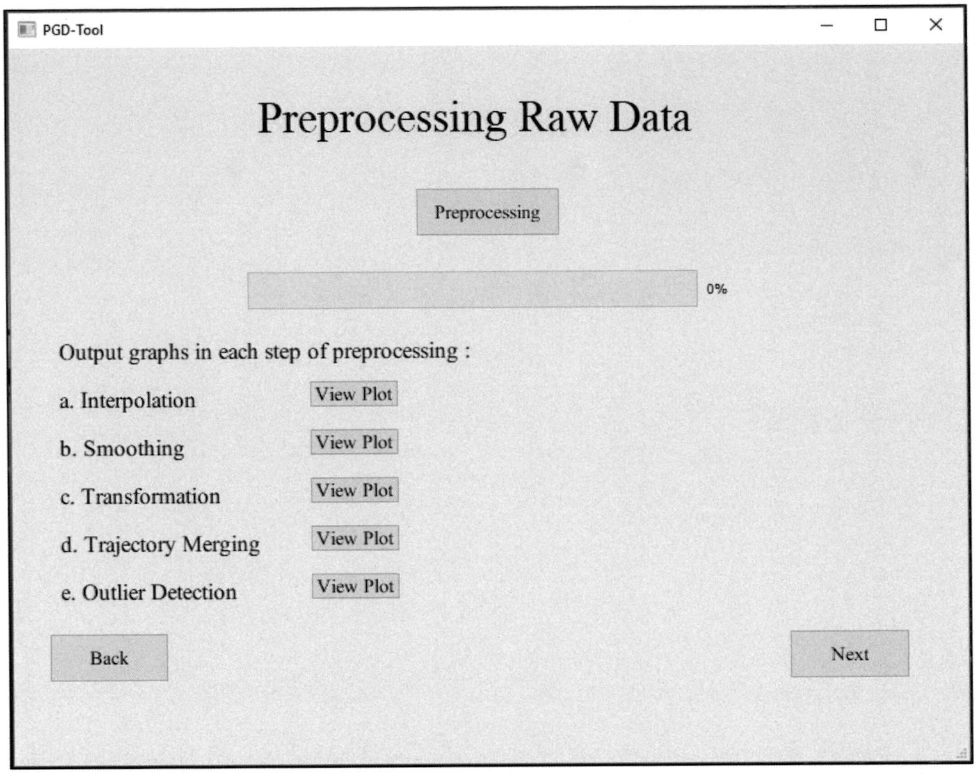

Figure 11.12 User Interface of page 3

Clicking on the "Preprocessing" button initiates all the functions in preprocessing. The progress bar below the button displays the status the preprocessing task. The "View Plot" button will display the graphs corresponding to each preprocessing step. Up to transformation preprocessing was performed on each Kinect individually. The x-axis of the graphs represents time frame number (or frame number), while the y-axis denotes the distance (in meters) covered by the subject while walking. The negative values (in y-axis) represent signed coordinate value for a particular Kinect. Each plot denotes the correspondence between the left and right ankles of a subject in anterior posterior (A-P) direction. The interpolation, smoothing, and transformation graphs (see Figs. 11.13–11.15, respectively) exhibit the operation performed on a single Kinect (i.e., 3rd Kinect). Trajectory merging (see Fig. 11.16) combines the data of all the Kinects and outlier detection (see Fig. 11.17) works on the combined data. The "Next" button remains inactive until the preprocessing is complete.

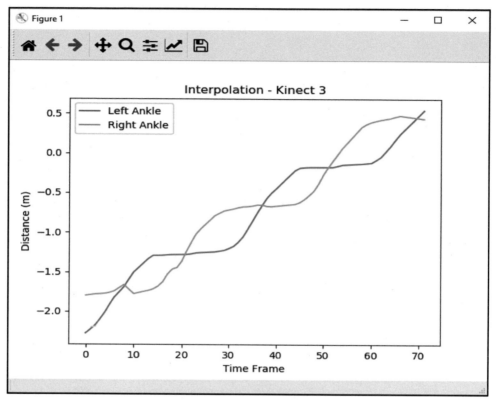

Figure 11.13 Interpolation.

The UI for page 4 of the tool is shown in Fig. 11.18. Performance of three machine learning techniques—LSTM, ANN, and SVM is shown in page 4. Clicking on any of the buttons will start the execution of the corresponding machine learning algorithm (Fig. 11.19).

Once the algorithm has completed its execution, a pop-up is displayed informing the same. Click on "OK" to view the results. Performance metrics accuracy, precision, sensitivity, and F1-score were used. After clicking on the button, "Comparative Analysis Chart," a bar chart will be displayed comparing the performance of the three classifiers. This button will become active only after all the algorithms have been executed at least once. It is to be noted that the application will run for the data acquisition architecture mentioned in Fig. 11.9.

11.4.3 Descriptions of the methods used

A brief description of the preprocessing methods used in this tool is given below:

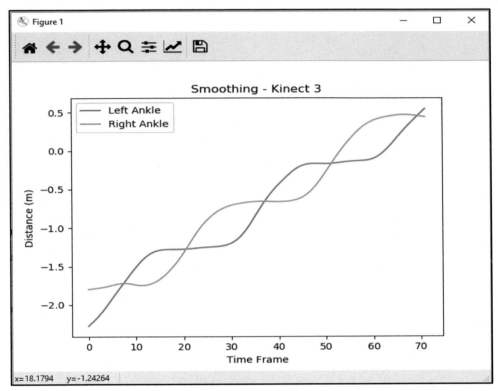

Figure 11.14 Smoothing.

- **Interpolation:** Sampling frequency of Kinect v2 sensor ranges from ~15 to ~30 Hz in practical cases. But, a constant sampling rate is favorable for extracting features from consecutive gait cycles. Hence, spline interpolation was used to make the sampling rate constant (i.e., 30 Hz) throughout all the gait cycles.
- **Smoothing:** Kinect v2 skeleton data stream suffers from high amount of noise due some undelaying problems (e.g., limited range of view, high sensitivity on view angle, etc.). Hence, to remove the noise, fourth order Butterworth low-pass filter was used.
- **Transformation:** For each joint, Kinect produce three-dimensional data. X and Y coordinates represent positional data in terms of pixel value, whereas, Z coordinate represents real distances between the joints and the camera. Hence, it is essential to first convert these coordinates into unified dimensions. This transformation is performed based on the knowledge that the real body length does not vary during walking and the procedure consists of the following steps:
  - Measuring the true height of the individual, denoted by H;

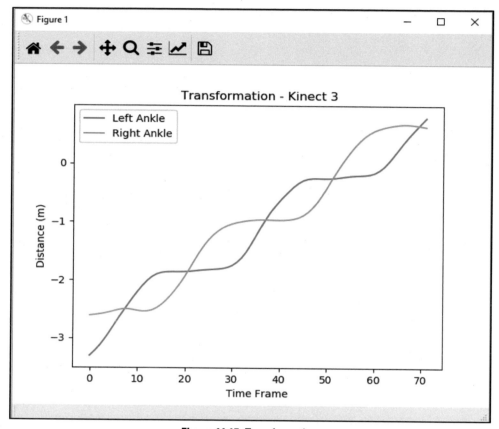

Figure 11.15 Transformation.

- Calculating the relative height based on the locations of the head and foot joints/points, denoted by h;
- Dividing H by h to obtain the transformation factor, denoted by R;
- Multiplying x and y by R to obtain their normalized values.
- **Trajectory Merging:** To get the entire walking path length, trajectories of individual Kinects were merged.
- **Outlier Detection and removal:** Despite removing high-frequency noise using low-pass filtering technique, some corrupted frames remain in the time series which need to detect and remove. A novel data-driven outlier detection and removal algorithm was used for that purpose. The algorithm constructs a sinusoidal time series by subtracting pelvis joint position from the corresponding ankle joints in A-P direction. Then from all maxima and minima, a threshold was constructed. If a set of frames between any consecutive extrema points is

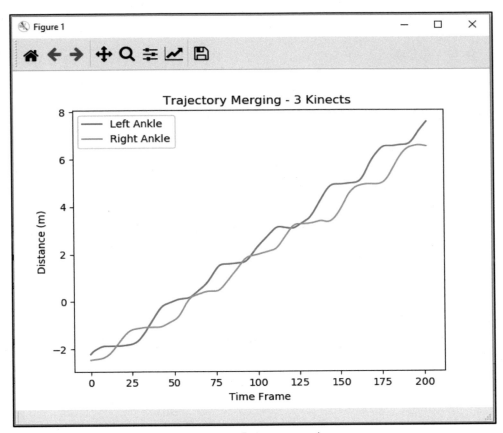

Figure 11.16 Trajectory merging.

below the said threshold, then all those frames were considered as outlier and subsequently removed from the time series. Then the gap was filled using interpolation.

For further description of the above-mentioned methods, see Refs. [3,4].

11.4.4 How to operate?

The tool that we have designed is a standalone tool. The user needs to install the tool in their computer. The tool can operate only in Windows OS (Windows 10 onwards). Once the tool is installed, the user the can use it as mentioned in the UI section.

11.4.5 Restrictions

There are very few limitations during the usage of this tool:
Restriction 1: The user can only upload ".zip" files to the tool.

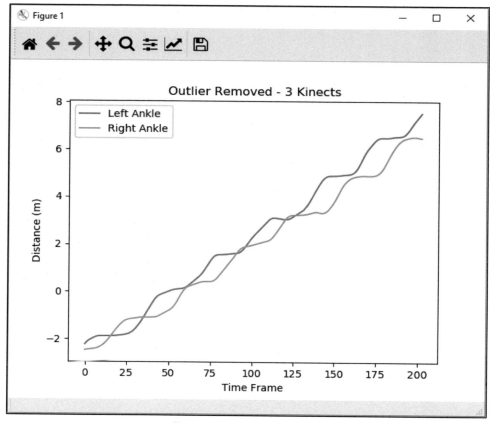

Figure 11.17 Outlier detection.

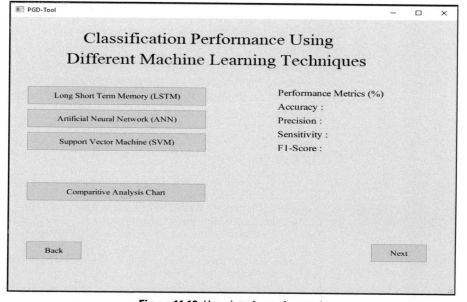

Figure 11.18 User interface of page 4

Figure 11.19 Pop-up message after clicking a button.

Restriction 2: Only ".csv" files should be present in the ".zip" file which was uploaded.

Restriction 3: During preprocessing and executing any algorithms, the user cannot click any other button.

Restriction 4: The user is not allowed to move from page 2 to page 3 unless uploading the input file is completed and from page 3 to page 4 unless preprocessing is completed.

Restriction 5: Presently, the tool is applicable only for the setup depicted in Fig. 11.9.

11.5 Summary

In this chapter, we discuss about the tools developed in our lab, "Machine Intelligence and Bio-motion Research Lab" in NIT Rourkela. These tools are developed on a graphical interactive platform for the researchers or clinical experts to obtain spatio-temporal features from gait signals and perform gait analysis based on that. The task of assessment of gait signals may become time-consuming and tiring for clinicians. Existing literature provides us ideas about processing the signal patterns automatically. We aggregated these ideas in a tool along with the option for user intervention. These tools may assist clinicians in the process of gait assessment.

References

[1] U. Martinez-Hernandez, A.A. Dehghani-Sanij, Adaptive Bayesian inference system for recognition of walking activities and prediction of gait events using wearable sensors, Neural Network. 102 (2018) 107−119.

[2] J. Chakraborty, A. Nandy, Periodicity detection of quasi-periodic slow-speed gait signal using IMU sensor, in: International Conference on Human-Computer Interaction, 2019, pp. 140−152.

[3] S. Chakraborty, N. Thomas, A. Nandy, Gait abnormality detection in people with cerebral palsy using an uncertainty-based state-space model, Lect.

Notes Comput. Sci. 12140 (Ml) (2020) 536−549. https://doi.org/10.1007/978-3-030-50423-6_40. LNCS.

[4]   S. Chakraborty, A. Nandy, Automatic diagnosis of cerebral palsy gait using computational intelligence techniques: a low-cost multi-sensor approach, IEEE Trans. Neural Syst. Rehabil. Eng. (2020). https://doi.org/10.1109/TNSRE.2020.3028203.

12

Conclusion

This book highlights the prospects of affordable gait analysis in the clinical domain. Existing systems for automated gait assessment are expensive and unaffordable for many clinics. High expenditure causes limited expansion of gait laboratories across the world. Only a few developed nations have managed to arrange the standard gait labs and clinics. Researches are also confined to some specific laboratories of universities and hospitals. In developing nations, like India, there are hardly a few gait labs. Compulsorily pathologists follow the error-prone qualitative techniques to assess gait. Hence, a low-cost arrangement to analyze gait is hugely needed. This book points out a set of affordable gait assessment systems, mainly for gait abnormality detection, gait event detection, different therapeutic interventions, and recovery prediction.

In the vision-based gait analysis, Kinect sensor-based systems have been described as an alternative to expensive cameras. This sensor is gradually gaining popularity in the clinical gait domain. Some studies have been pointed out which have validated different data streams of this sensor. State-of-the-art studies have obtained a competing result while using this sensor for different clinical issues of gait. This book also provides a step-by-step guideline to construct a Kinect-based gait abnormality detection system. The recommended system can also be used in other problems related to clinical gait like event detection, different therapeutic interventions, etc. In addition, some statistical techniques, which are popular in this domain are also highlighted. However, some challenges remain there, like view angle of Kinect, limited range of depth-sensing, noise in the skeletal data stream, etc., which limited its usage to some extent.

Wearable sensors such as inertial sensors have the advantages of being economically viable and portable. Inertial sensor holds the potential to replace highly expensive motion capture systems to obtain kinematic information. However, it is still not used for clinical gait labs in practical because of the lack of accuracy and

Modern Methods for Affordable Clinical Gait Analysis. https://doi.org/10.1016/B978-0-323-85245-6.00001-1

precision of gait analysis systems with inertial sensor devices. Various validation studies which reported the comparable accuracy of inertial measurement unit (IMU)-based systems have been presented. The research on clinical gait analysis has revealed a number of possible solutions to develop reliable technologies to diagnose and monitor gait abnormality in an early stage using artificial intelligence and machine learning techniques. We have tried to address various essential processes and methodologies for assessment of gait signals obtained from inertial sensors. We highlighted the state-of-the-art techniques that have been proposed for pathological gait pattern analysis for healthcare applications. One of the foremost challenges of using low-cost sensor is the extraction of salient features from highly noisy data to characterize a particular gait pattern. For this purpose, we discussed the techniques for IMU sensor calibration techniques which can help in obtaining corrected signal values, gait signal segmentation techniques that can help in extracting different types of features. Apart from identification of pathological condition, other therapeutic intervention aspects of inertial sensors such as musculoskeletal injury prevention and recovery are also discussed. Although having potential to be a competent tool for gait analysis, usage of wearable sensors in clinical gait analysis has some limitations that still need attention, e.g., automated calibration techniques that requires no manual intervention, lighter sensors that are less intrusive to the users, etc.

EMG sensors are used to capture the activity pattern of muscles which are responsible for any movement made by a human being. The activity will be considered as the effect of activity in the muscles. The EMG signals offer the muscle activation pattern during various motor action. The myoelectric signals demonstrate how the muscle actions have an impact on the mechanical movement. The amplitude of signals is proportionate to the relative muscle tension. Different types of gait activities are identified using surface electromyography (EMG) signals. Various time and frequency domain features are extracted from each cycle. After feature extraction, different classifiers are used for classification on the basis of the data collected from the lower limb muscle. It can be concluded that usage of sEMG along with other wearable sensors like IMU sensors would be beneficial in clinical gait analysis. This book discusses that the sensors such as EMG can be used for automated monitoring of activities. Apart from this, it also provides vital suggestions for other EMG signal-based devices, such as clinical applications, walking assist devices, and robotics or prosthetic gadgets.

We recommend a set of affordable gait assessment systems as an alternative to the prevailing high-cost systems. The book high-lights the usage of Kinect as a vision sensor and wearable sensors such as miniature IMU and surface electromyography.

Index

Note: Page numbers followed by "f" indicate figures and "t" indicate tables.

Printed in the United States
by Baker & Taylor Publisher Services